应用型本科 机械类专业"十三五"规划教材

机械工程专业英语

主　编　肖　平　韩利敏

副主编　秦广虎　王　娟　业红玲　牛礼民　罗玉枝

西安电子科技大学出版社

内容简介

本书内容根据机械工程领域的知识特点分为机械工程材料、材料的成型方法、机械零部件、机械加工设备、工程机械设备、电气知识及汽车等六大部分。全书共30个单元，每个单元均包括课文和阅读材料两大部分，课文和阅读材料之后给出了词汇、注释、思考题等内容。本书内容的编写顺序与机械工程专业培养方案基本一致，这有助于读者阅读与学习。

本书可作为高等工科院校大机械类相关专业的英语教材，也可作为机械工程及相关专业的管理人员和技术人员自学参考用书。

图书在版编目(CIP)数据

机械工程专业英语/肖平，韩利敏主编. —西安：西安电子科技大学出版社，2017.6
应用型本科 机械类专业"十三五"规划教材
ISBN 978—7—5606—4496—7

Ⅰ.① 机⋯ Ⅱ.① 肖⋯ ② 韩⋯ Ⅲ.① 机械工程—英语 Ⅳ.① TH

中国版本图书馆 CIP 数据核字(2017)第 104731 号

策　　划　高　樱
责任编辑　卢　杨　阎　彬
出版发行　西安电子科技大学出版社(西安市太白南路2号)
电　　话　(029)88242885　88201467　　邮　编　710071
网　　址　www.xduph.com　　　　　　　电子邮箱　xdupfxb001@163.com
经　　销　新华书店
印刷单位　陕西华沐印刷科技有限责任公司
版　　次　2017年6月第1版　　2017年6月第1次印刷
开　　本　787毫米×1092毫米 1/16　印 张　15
字　　数　353千字
印　　数　1～3000册
定　　价　30.00元

ISBN 978—7—5606—4496—7/TH

XDUP 4788001-1

如有印装问题可调换

应用型本科 机械类专业规划教材
编审专家委员会名单

主　任：张　杰（南京工程学院 机械工程学院 院长/教授）
副主任：杨龙兴（江苏理工学院 机械工程学院 院长/教授）
　　　　张晓东（皖西学院 机电学院 院长/教授）
　　　　陈　南（三江学院 机械学院 院长/教授）
　　　　花国然（南通大学 机械工程学院 副院长/教授）
　　　　杨　莉（常熟理工学院 机械工程学院 副院长/教授）
成　员：（按姓氏拼音排列）
　　　　陈劲松（淮海工学院 机械学院 副院长/副教授）
　　　　高　荣（淮阴工学院 机械工程学院 副院长/教授）
　　　　郭兰中（常熟理工学院 机械工程学院 院长/教授）
　　　　胡爱萍（常州大学 机械工程学院 副院长/教授）
　　　　刘春节（常州工学院 机电工程学院 副院长/副教授）
　　　　刘　平（上海第二工业大学 机电工程学院 教授）
　　　　茅　健（上海工程技术大学 机械工程学院 副院长/副教授）
　　　　唐友亮（宿迁学院 机电工程系 副主任/副教授）
　　　　王荣林（南理工泰州科技学院 机械工程学院 副院长/副教授）
　　　　王树臣（徐州工程学院 机电工程学院 副院长/教授）
　　　　王书林（南京工程学院 汽车与轨道交通学院 副院长/副教授）
　　　　吴懋亮（上海电力学院 能源与机械工程学院 副院长/副教授）
　　　　吴　雁（上海应用技术学院 机械工程学院 副院长/副教授）
　　　　许德章（安徽工程大学 机械与汽车工程学院 院长/教授）
　　　　许泽银（合肥学院 机械工程系 主任/副教授）
　　　　周　海（盐城工学院 机械工程学院 院长/教授）
　　　　周扩建（金陵科技学院 机电工程学院 副院长/副教授）
　　　　朱龙英（盐城工学院 汽车工程学院 院长/教授）
　　　　朱协彬（安徽工程大学 机械与汽车工程学院 副院长/教授）

前　　言

随着经济社会的迅速发展，我国已成为名副其实的制造业大国。国内的机械产品逐渐走向世界，同时世界著名的机械产品制造商也纷纷加入中国市场。这种技术大融合的时代背景和产业全球化的趋势必然要求培养大量既掌握机械领域专业知识又具备良好英语沟通能力的高级人才。为了使安徽省高等教育赶上并顺应时代发展的步伐，我们邀请了安徽省内一些机械工程专业较强的高校共同编写了这本《机械工程专业英语》。本书可作为大专院校机械工程专业的英语授课教材，也可以作为机械行业工程技术人员继续教育和职工岗位培训的教材，还可供专业技术人员熟悉专业词汇和学习机械工程专业英语使用。

本书共分为六大部分，每个部分分为五个单元进行介绍。第一部分为机械工程材料，主要讲述机械工程材料的类型及特点；第二部分为机械工程材料的成型方法，主要讲述机械工程材料的浇注、冲压等成型方法；第三部分为机械零部件，主要讲述机械工程领域中常用的零部件及其工作原理；第四部分为机械加工设备，主要介绍车床、铣床等常用加工设备；第五部分为工程机械设备，主要介绍挖土机、起重机等常用工程机械设备；第六部分为电气知识及汽车，主要介绍机械工程领域常用的电气设备原理及汽车领域相关知识。书中内容选自国外的专业书刊和大专院校教科书的部分章节，语言简明流畅，难易适中，实用性较强。通过学习本书，读者不仅可以提高英语水平，而且可以熟悉机械工程方面的专业知识。在编写本书的过程中，考虑到我国读者的英文水平，我们尽可能使每篇课文的篇幅适度，难度适中；为帮助读者克服阅读障碍，每篇课文后有注释和词汇表，课文中出现的长难句和语法难点均在课后的注释中加以解释。同时，本书的编写顺序与汽车类教材基本一致，这也有助于广大读者的阅读与理解。

我们编写本书的目的是希望读者通过学习本书，可以提高对机械工程专业英文技术资料的理解能力以及对机械工程专业技术正确进行英文表达的能力。这就需要读者在学习和阅读本书的过程中注重对本书专业词汇及固定英文表达方式、方法的掌握。

本书由安徽工程大学肖平(第1～第10单元)、韩利敏(第11～第20单元)担任主编；嘉兴学院王娟(第21、22、23单元)、安徽工程大学秦广虎(第24、25单元)、安徽农业大学罗玉枝(第26单元)、蚌埠学院业红玲(第27、28单元)、安徽工业大学牛礼民(第29、30单元)担任副主编；安徽工程大学高洪教授担任主审。参加本书编写的人员还有安徽工程大学车辆工程教研室的时培成、王刚、潘道远、唐冶、周加冬、王海涛等老师，以及合肥工业大学夏锡全、安徽工业大学王永宽、安徽科技学院易克传、安徽农业大学张小龙、安徽工程大学外国语学院聂珺同学及硕士研究生聂高法等。

在本书编写过程中，我们参阅了大量的书籍和文献资料，受益匪浅，在此向有关作者表示衷心的感谢！

由于编者水平有限，加上时间紧迫，书中疏漏之处在所难免，敬请广大专家和读者批评指正。

编　者

2017年2月

Contents

PART 1   Materials

Unit 1   Engineering Materials 2
    Reading Material: Metallic and Nonmetallic Materials 4
Unit 2   Mechanical Properties of Metals 6
    Reading Material: Stress and Strain 9
Unit 3   Steels 14
    Reading Material: Selection of Construction Materials 17
Unit 4   Cast Iron 21
    Reading Material: Metals and Their Use 23
Unit 5   Stainless Steels 25
    Reading Material: Selection of Stainless Steel 27

PART 2   Forming Technology

Unit 6   Casting 30
    Reading Material: Ingot Casting 32
Unit 7   Forging 35
    Reading Material: Forging 38
Unit 8   Heat Treatment of Metals 41
    Reading Material: Heat Treatment of Steels 48
Unit 9   Welding 52
    Reading Material: Welding 56
Unit 10   Die Casting 62
    Reading Material: Sand Casting 64

PART 3   Mechanical Elements

Unit 11   Gears 72
    Reading Material: Spur Gears 74
Unit 12   Shafts and Couplings 78
    Reading Material: Couplings 81
Unit 13   Clutches 86
    Reading Material: Shafts, Clutches and Brakes 88
Unit 14   Rolling Contact Bearings 92
    Reading Material: Bearings 94
Unit 15   Belts, Clutches, Brakes, and Chains 97

Reading Material: Worm Gear Sets 100

PART 4 Machining Tools

Unit 16 Lathes 104
    Reading Material: Machine Tools 106
Unit 17 Drills and Drilling Machines 109
    Reading Material: Radial Drilling Machine 112
Unit 18 Milling, Shaper, Planer and Grinding Machines 113
    Reading Material: Milling Machines and Grinding Machines 114
Unit 19 Machine Tool Tests, Accuracy Checking and Maintenance 119
    Reading Material: Physical Basis of the Cutting Process 122
Unit 20 Nontraditional Manufacturing Processes 128
    Reading Material: Cutting Forces and Cutting Power 131

PART 5 Engineering Machines

Unit 21 The Two Luffing Cable Cranes 138
    Reading Material：Hydraulic System 143
Unit 22 Komatsu D375A Bulldozer 147
    Reading Material: Lubrication 152
Unit 23 Kawasaki Power Loader 156
    Reading Material: Computer Technology 160
Unit 24 Komatsu Truck 168
    Reading Material: Automobile Components 172
Unit 25 CAD/CAM/CAPP 178
    Reading Material: Components of a Robot System 182

PART 6 Electric Knowledge and Automobile

Unit 26 Numerical Control of Production Equipments (I) 186
    Reading Material: Numerical Control of Production Equipments (II) 191
Unit 27 Basic Electricity and Magnetism 199
    Reading Material: Direct-current Circuits 202
Unit 28 Electric Powers 206
    Reading Material: Electrical Instruments and Electrical Measurements 210
Unit 29 Numerical Control Software 213
    Reading Material: Transducers and Sensors 216
Unit 30 Automatic Control System 222
    Reading Material: Artificial Intelligence for Automotive Manufacturing 228

参考文献 231

PART 1 Materials

Unit 1 Engineering Materials

All products that come out of industry consist of at least one and often many types of materials. The most obvious example is the automobile. A car contains a wide variety of materials, ranging from glass to steel to rubber, plus numerous other metals and plastics.

The number of materials which are available to the engineer in industry is almost infinite. The various compositions of steel alone run into the thousands. It has been said that there are more than 10,000 varieties of glass, and the numbers of plastics are equally great. It addition, several hundred new varieties of materials appear on the market each month. This means that individual engineers and technicians cannot hope to be familiar with all the properties of all types of materials in their numerous forms. All that he can do is try to learn some principles to guide him in the selection and processing of materials.

The properties of a material originate from the internal structure of that material. This is analogous to saying that the operation of a TV set depends on the components and circuits within that set. The internal structures of materials involve atoms, and the way atoms are associated with their neighbors into crystals, molecules, and microstructures.

It is convenient to divide materials into three main types: (1) metals (2) plastics or polymers and (3) ceramics.

Characteristically, metals are opaque, ductile, and good conductors of heat and electricity. Plastics (or polymers), which usually contain light elements and have relatively low density, are generally insulators, and are flexible and formable at relatively low temperatures. Ceramics, which contain compounds of both metallic and nonmetallic elements, are usually relatively resistant to severe mechanical, thermal, and chemical conditions.

Metals are divided into ferrous and non-ferrous metals. The former contain iron and the latter do not contain iron. Certain elements can improve the properties of steel and are therefore added to it. For example, chromium may be included to resist corrosion and tungsten to increase hardness. Aluminum, copper, and the alloys, bronze and brass, are common non-ferrous metals.

Plastics and ceramics are non-metals; however, plastics may be machined like metals. Plastics are classified into two types: thermoplastics and thermosets. Thermoplastics can be shaped and reshaped by heat and pressure but thermosets cannot be reshaped because they undergo chemical changes as they harden. Ceramics are often employed by engineers when materials which can withstand high temperatures are needed.

◇ New Words and Expressions

variety [vəˈraɪəti]	n. 多样性
infinite [ˈɪnfɪnət]	adj. 无限的
composition [ˌkɒmpəˈzɪʃn]	n. 合成 (物)
property [ˈprɒpəti]	n. 性能
principle [ˈprɪnsəpl]	n. 规则
originate [əˈrɪdʒɪneɪt]	vi. 起源
internal [ɪnˈtɜːnl]	adj. 内部的
analogous [əˈnæləgəs]	adj. 相似的
component [kəmˈpəʊnənt]	n. 一个部分
crystal [ˈkrɪstl]	n. 晶体，石英
molecule [ˈmɒlɪkjuːl]	n. 分子，微小颗粒
microstructure [ˈmaɪkrəʊˌstrʌktʃə]	n. 微结构
convenient [kənˈviːniənt]	adj. 方便的
polymer [ˈpɒlɪmə(r)]	n. 聚合物
ceramics [sɪˈræmɪks]	n. 陶瓷
characteristically [ˌkærəktəˈrɪstɪkli]	adv. 典型地
opaque [əʊˈpeɪk]	adj. 模糊的
ductile [ˈdʌktaɪl]	adj. 有韧性的，易延展的
conductor [kənˈdʌktə(r)]	n. 导体
insulator [ˈɪnsjuleɪtə(r)]	n. 绝缘体
flexible [ˈfleksəbl]	adj. 易弯曲的
formable [ˈfɔːməbl]	adj. 易成型的
compound [ˈkɒmpaʊnd]	n. 化合物
metallic [məˈtælɪk]	adj. 金属的
ferrous [ˈferəs]	adj. 亚铁的，含铁的
chromium [ˈkrəʊmiəm]	n. 铬
corrosion [kəˈrəʊʒn]	n. 腐蚀
tungsten [ˈtʌŋstən]	n. 钨
aluminum [əˈljuːmɪnəm]	n. 铝
consist of	由……组成
range from...to...	从……至……
run into	多达
associate with	与……相关
the former...,the latter...	前者……，后者……

▷ Questions

1. What materials does a car contain?
2. Why do individual engineers and technicians cannot be familiar with all types of materials of cars?
3. How can the materials of cars be divided?
4. What are the main features of metals, plastics and ceramics respectively?
5. When are ceramics often employed by engineers?

Reading Material: Metallic and Nonmetallic Materials

Perhaps the most common classification that is encountered in materials selection is whether the material is metallic or nonmetallic. The common metallic materials are such metals as iron, copper, aluminum, magnesium, nickel, titanium, lead, tin, and zinc and the alloys of these metals, such as steel, brass, and bronze. They possess the metallic properties of luster, thermal conductivity, and electrical conductivity; are relatively ductile; and some have good magnetic properties. The common nonmetals are wood, brick, concrete, glass, rubber, and plastics. Their properties are very wide, but they generally tend to be less ductile, weaker, and less dense than the metals, and they have no electrical conductivity and poor thermal conductivity.

Although it is likely that metals always will be the more important of the two groups, the relative importance of the nonmetallic group is increasing rapidly, and since new nonmetals are being created almost continuously, this trend is certain to continue. In many cases the selection between a metal and nonmetal is determined by a consideration of required properties. Where the required properties are available in both, total cost becomes the determining factor.

◇ New Words and Expressions

encounter [ɛnˈkaʊntɚ]	v. 碰见
bronze [brɑːnz]	n. 青铜
luster [ˈlʌstɚ]	n. 光泽
magnesium [mægˈniːziəm]	n. 镁
nickel [ˈnɪkəl]	n. 镍
titanium [tɪˈteɪniəm]	n. 钛
It is likely that…	很可能
It is certain to	必然，一定
tend to	趋向于

⊠ Questions

1. Can you give some examples of common metallic materials?
2. Can you give some examples of common nonmetallic materials?
3. What properties do metallic materials have?
4. What properties do nonmetals have?
5. What determines the selection between a metal and a nonmetal?

Unit 2 Mechanical Properties of Metals

Mechanical properties are the characteristic responses of a material to applied forces. These properties fall into five broad categories: strength, hardness, elasticity, ductility, and toughness.

Strength is the ability of a metal to resist applied forces.

Strength properties are commonly referred to as tensile strength, bending strength, compressive strength, torsional strength, shear strength and fatigue strength.

Tensile strength is that property which resists forces acting to pull the metal apart[1]. It is one of the most important factors in the evaluation of a metal.

Compressive strength is the ability of a material to resist being crushed. Compression is the opposite of tension with respect to the direction of the applied load[2]. Most metals have high tensile strength and high compressive strength. However, brittle materials such as cast iron have high compressive strength but only a moderate tensile strength.

Bending strength is that quality which resists forces from causing a member to bend or deflect in the direction in which the load is applied. Actually a bending stress is a combination of tensile and compressive stresses.

Torsional strength is the ability of a metal to withstand forces that cause a member to twist.

Shear strength refers to how well a member can withstand two equal forces acting in opposite directions.

Fatigue strength is the property of a material to resist various kinds of rapidly alternating stresses. For example, a piston rod or an axle undergoes complete reversal of stresses from tension to compression. Bending a piece of wire back and forth until it breaks is another example of fatigue strength.

Hardness is that property in steel which resists indentation or penetration. Hardness is usually expressed in terms of the area of an indentation made by a special ball under a standard load, or the depth of a special indenter under a specific load[3].

Elasticity is the ability to spring back to original shape. Auto bumpers and all springs should have this quality.

Ductility is the ability to undergo permanent changes of shape without rupturing. Modern, deep-formed auto bodies and fenders, and other stamped and formed products must have this property.

Toughness is the ability to absorb mechanically applied energy. Strength and ductility determine a material's toughness. Toughness is needed in railroad cars, automobile axles, hammers, rails, and similar products.

◇ *New Words and Expressions*

mechanical [mɪ'kænɪk(ə)l]	adj. 力学的，机械的
response [rɪ'spɒns]	n. 反应，响应，答复
broad [brɔːd]	adj. 主要的，概括的
category ['kætɪg(ə)rɪ]	n. 种类，类别
strength [streŋθ; streŋkθ]	n. 强度
hardness ['hɑːdnəs]	n. 硬度
elasticity [elæ'stɪsɪtɪ]	n. 弹性，弹力
toughness ['tʌfnis]	n. 韧性
tensile ['tensaɪl]	adj. 拉伸的
compressive [kəm'presɪv]	adj. 压缩的
torsional ['tɔːʃənəl]	adj. 扭转的
shear [ʃɪə]	n. 剪切
fatigue [fə'tiːg]	n. 疲劳
evaluation [ɪˌvælju'eɪʃn]	n. 评价，评定
crush [krʌʃ]	v. 压碎
tension ['tenʃ(ə)n]	n. 拉伸，绷紧
brittle ['brɪt(ə)l]	adj. 脆的
cast iron	铸铁
moderate ['mɒd(ə)rət]	adj. 适度的，温和的，中等的
member ['membə]	n. 构件，零件
deflect [dɪ'flekt]	v. 弯曲，挠曲
stress [stres]	n. 应力
twist [twɪst]	v. 扭转，捻
withstand [wɪð'stænd]	v. 抵抗，经受住
alternating ['ɔːltəneɪtɪŋ]	adj. 交变的
piston ['pɪst(ə)n]	n. 活塞，柱塞
axle ['æks(ə)l]	n. 车轴，轮轴，心棒
undergo [ʌndə'gəʊ]	vt. 经历，经受
reversal [rɪ'vɜːs(ə)l]	n. 变换，颠倒，改变符号
indentation [ɪnden'teɪʃ(ə)n]	n. 压痕，凹痕
indenter [ɪn'dentə]	n. (硬度试验的) 压头
penetration [penɪ'treɪʃ(ə)n]	n. 穿透，锥入程度

rupture ['rʌptʃə]	v. 断裂
deep-formed	adj. 深成型的
auto ['ɔːtəʊ]	n. 汽车；adj. 自动的
bumper ['bʌmpə]	n. 保险杠，缓冲器
fender ['fendə]	n. 挡板，防护板
stamp [stæmp]	vt. 冲压，模压
fall into	分为
be referred to as	成为，叫做
with respect to	就……而论，相对于
in terms of	依据，根据

⇨ Notes

[1] Tensile strength is that property which resists forces acting to pull the metal apart.

参考译文：抗拉强度是金属抵抗外力把它拉断的能力。

句中关系代词 which 引导一个定语从句修饰 property，在从句中 which 作主语。此外，在该从句中，现在分词短语 acting to pull the metal apart 作定语修饰 forces。Pull…apart 意为"把……拉开，把……拉断"。

[2] Compression is the opposite of tension with respect to the direction of the applied load.

参考译文：压缩时载荷的方向与拉伸时的载荷方向相反。

该句中介词短语 with respect to the direction of the applied load 作全句的状语。

[3] Hardness is usually expressed in terms of the area of an indentation made by a special ball under a standard load, or the depth of a special indenter under a specific load.

参考译文：硬度通常用在标准载荷作用下特制球产生的压痕面积来表示，或用特定载荷下专门压头所形成的深度来表示。

句中 the area…a standard load 和 the depth…a specific load 为并列的两个名词性短语，通过等立连词 or 连接，它们都是短语介词 in terms of 的宾语。介词短语 in terms of…a specific load 作全句的状语。

⇨ Questions

1. Which categories do the mechanical properties mainly fall into?
2. What is the definition of tensile strength?
3. What is hardness?
4. What is ductility?
5. What is toughness?

Reading Material: Stress and Strain

The fundamental concepts of stress and strain can be illustrated by considering a prismatic bar that is loaded by axial forces F at the ends, as shown in Figure 2-1. A prismatic bar is a straight structural member having constant cross section throughout its length. In this illustration, the axial forces produce a uniform stretching of the bar, hence, the bar is said to be in tension.

To investigate the internal stresses produced in the bar by the axial forces, we make an imaginary cut at section mn (Figure 2-1(a)).

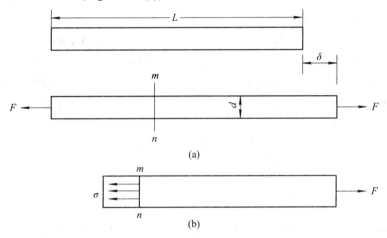

Figure 2-1 Prismatic bar in tension

This section is taken perpendicular to the longitudinal axis of the bar; hence, it is known as a cross section. We now isolate the part of the bar to the right of the cut as a free body (Figure2-1(b)). The tensile load F acts at the right end of the free body; at the other end are forces representing the action of removed part of the bar upon the part that remains. These forces are continuously distributed over the cross section, analogous to the continuous distribution of hydrostatic pressure over a submerged horizontal surface. The intensity of force (that is, the force per unit area) is called the stress and is commonly denoted by the Greek letter σ (sigma). Assuming that the stress has a uniform distribution over the cross section(see Figure 2-1(b)), we can readily see that its resultant is equal to the intensity σ times the cross-sectional area A of the bar. Furthermore, from the equilibrium of the body shown in Figure 2-1(b), it is also evident that this resultant must be equal in magnitude and opposite in direction to the applied load F. Hence, we obtain

$$\sigma = \frac{F}{A} \tag{2-1}$$

As the equation for the uniform stress in an axially loaded, prismatic bar of arbitrary cross-sectional shape. When the bar is stretched by the forces F, as shown in the figure, the resulting stresses are tensile stresses; if the forces are reversed in direction, causing the bar to be compressed, we obtain compressive stresses. In as much as the stress σ acts in a direction perpendicular to the cut surface, it is referred to as a normal stress. Thus, normal stresses may be either tensile or compressive stresses.

When a sign convention for normal stresses is required, it is customary to define tensile stresses as positive and compressive stresses as negative.

Because the normal stress σ is obtained by dividing the axial force by the cross-sectional area, it has units of force per unit of area. When SI units are used, force is expressed in newtons (N) and area in square meters (m^2)[1]. Hence, stress has units of newtons per square meter (N/m^2), or pascals (Pa). However, the pascal is such a small unit of stress that it is necessary to work with large multiples. To illustrate this point, we have only to note that it takes almost 7,000 Pascals to make 1 per square inch (psi). As an example, a typical tensile stress in a steel bar might have a magnitude of 140 megapascals (140MPa), which is 140×10^6 pascals. Other units that may be convenient to use are the kilopascal (kPa) and gigapascal (GPa); the former equals 10^3 pascals and the latter equals 10^9 pascals. Although it is not recommended in SI, you will sometimes find stress given in newtons per square millimeter (N/mm^2), which is a unit identical to the megapascal (MPa).

When using USCS units, stress is customarily expressed in pounds per square inch (psi) or kips per square inch (ksi) [2]. For instance, a typical stress in a steel bar might be 20,000 psi or 20 ksi.

In order for the equation $\sigma = F/A$ to be valid, the stress σ must be uniformly distributed over the cross section of the bar. This condition is realized if the axial force F acts through the centroid of the cross sectional area, as demonstrated in Example 1. When the load F does not act at the centroid, bending of the bar will result, and a more complicated analysis is necessary. However, we will assume throughout this book that all axial forces are applied at the centroid of the cross section unless specifically stated otherwise.

The uniform stress condition pictured in Figure 2-1(b) exists throughout the length of the member except near the ends. The stress distribution at the ends of the bar depends upon the details of how the axial load F is actually applied. If the load itself is distributed uniformly over the end, then the stress pattern at the end will be the same as elsewhere. However, the load is usually concentrated over a small area, resulting in high localized stresses and nonuniform stress distributions over cross sections in the vicinity of the load. As we move away from the ends, the stress distribution gradually approaches the uniform distribution shown in Figure 2-1(b). It is usually safe to assume that the formula $\sigma = F/A$ may be used with good accuracy at any point within the bar that is at least a distance d away from the ends, where d is the largest transverse dimension of the bar (see Figure 2-1(a))[3]. Of course, even when the stress is not uniform, the equation $\sigma = F/A$ will give the average normal stress.

An axially loaded bar undergoes a change in length, becoming longer when in tension and shorter when in compression. The total change in length is denoted by the Greek letter δ (delta) and is pictured in Figure 2-1(a) for a bar in tension. This elongation is the cumulative result of the stretching of the material throughout the length L of the bar. Let us now assume that the material is the same everywhere in the bar. Then, if we consider the half of the bar, it will have an elongation equal to $\delta/2$; similarly, if we consider a unit of the bar, it will have an elongation equal to $1/L$ times the total elongation δ. In this manner, we arrive at the concept of elongation per unit length, or strain, denoted by the Greek letter ε (epsilon) and given by the equation.

$$\varepsilon = \frac{\delta}{L} \qquad (2\text{-}2)$$

If the bar is in tension, the strain is called a tensile strain, representing an elongation or stretching of the material. If the bar is in compression, the strain is a compressive strain and the bar shortens. Tensile strain is taken as positive, and compression strain as negative. The strain ε is called a normal strain because it is associated with normal stresses.

Because normal strain ε is the ratio of two lengths, it is a dimensionless quantity; that is, it has no units. Thus, strain is expressed as a pure number, independent of any system of unit. Numerical values of strain are usually very small, especially for structural materials, which ordinarily undergo only small changes in dimensions. As an example, consider a steel bar having length L of 2.0 m. When loaded in tension, the bar might elongate by an amount δ equal to 1.4mm. The corresponding strain is

$$\varepsilon = \frac{\delta}{L} = \frac{1.4 \times 10^{-3}}{2.0\text{m}} = 0.0007 = 700 \times 10^{-6}$$

In practice, the original units of Q and L are sometimes attached to the strain itself, and then the strain is recorded in forms such as mm/m, μm/m, and in./in. For instance, the strain in the preceding illustration could be given as 700μm/m or 700×10^{-6} in./in.

The definitions of normal stress and strain are based upon purely statical and geometrical considerations, hence Eqs. (2-1) and (2-2) can be used for loads of any magnitude and for any material. The principal requirement is that the deformation of the bar be uniform, which in turn requires that the bar be prismatic, the loads act through the centroids of the cross sections, and the material be homogeneous (that is, the same throughout all parts of the bar)[4]. The resulting state of stress and strain is called uniaxial stress and strain.

◇ *New Words and Expressions*

stress [stres]	n. 应力
strain [strein]	n. 应变
prismatic [prɪz'mætɪk]	adj. 等截面的
axial ['æksɪəl]	adj. 轴（向）的
tension ['tenʃ(ə)n]	n. 拉力，拉伸
perpendicular [ˌpɜːp(ə)n'dɪkjʊlə]	adj. 与……垂直(正交)的
longitudinal [lɒndʒə'tudnl]	adj. 纵(向)的，轴向的

tensile ['tensaɪl]	adj. 拉伸的
analogous [ə'næləgəs]	adj. 类比的，模拟的
hydrostatic [haɪdrə(ʊ)'stætɪk]	adj. 液压(水)静力的
submerge [səb'mɜːdʒ]	v. 浸(沉)没
denote [dɪ'nəʊt]	vt. 指(表)示，代表
resultant [rɪ'zʌlt(ə)nt]	n. 结果，合力
times [taɪmz]	prep. 乘
equilibrium [ˌikwɪ'lɪbrɪəm]	n. 平衡，稳定
magnitude ['mægnɪtjuːd]	n. 数量，数值
normal ['nɔːm(ə)l]	adj. &n. 垂直(的)
arbitrary ['ɑːbɪt(rə)rɪ]	adj. 随机的
centroid ['sentrɒɪd]	n. 矩心，重心
vicinity [vɪ'sɪnɪtɪ]	n. 附近，周围
transverse [træns'vɜ·s]	adj. 横(向)的
elongation [iːlɒŋ'geɪʃ(ə)n]	n. 拉长，伸长(率)
cumulative ['kjuːmjʊlətɪv]	adj. 累积的
dimensionless [daɪ'mɛnʃənlɪs]	adj. 无单位的，无量纲的
quantity ['kwɒntɪtɪ]	n. 参数，值
precede [prɪ'siːd]	vt. 在……之前，先于
deformation [ˌdiːfɔː'meɪʃ(ə)n]	n. 变形
homogeneous [ˌhomə'dʒɪnɪəs]	adj. 均匀的，均质的
uniaxial [juːnɪ'æksɪəl]	adj. 单轴(向)的，同轴的
cross section	横截面
tensile stress	拉伸应力
compressive stress	压缩应力
normal stress	正应力
inasmuch as	因为，由于
USCS=United States Customary System	美国单位制
tensile strain	拉应变
compressive strain	压应变
normal strain	正应变

⇨ *Notes*

[1] When SI units are used, force is expressed in newtons (N) and area in square meters(m^2).
参考译文：采用国际单位制时，力用牛顿(N)表示，面积用平方米(m^2)表示。
SI 是国际单位制的简称，其全称是 Systeme International d' Unites，这是法语。

[2] When using USCS units, stress is customarily expressed in pounds per square inch (psi)

or kips per square inch(ksi).

参考译文：当用美国单位制时，应力通常用磅/平方英寸(psi)或千磅/平方英寸(ksi)表示。USCS 是美国单位制的简称，其全称是 United States Customary System。

[3] It is usually safe to assume that the formula $\sigma = F/A$ may be used with good accuracy at any point within the bar that is at least a distance d away from the ends, where d is the largest transverse dimension of the bar (see Figure 2-1(a)).

参考译文：假设公式 $\sigma = F/A$ 能较好地适用于杆件内部离开两端距离至少不小于 d 的地方，这通常是安全的，此处 d 为杆的最大横向尺寸(见图 2-1(a))。

(1) 本句中，it 是形式主语，真正的主语是不定式短语 to assume...the bar(see Figure 2-1(a))。

(2) 在主语中，第一个 that 引导的是动词 assume 的宾语从句，第二个 that 引导的是定语从句修饰名词 point，介词短语 within the bar 作定语修饰 point。

(3) where d is the largest transverse dimension of the bar 作状语从句，对全句作补充说明。

[4] The principal requirement is that the deformation of the bar be uniform, which in turn requires that the bar be prismatic, the loads act through the centroids of the cross sections, and the material be homogeneous (that is, the same throughout all parts of the bar).

参考译文：主要的要求是棒的变形是均匀的，同时要求棒料是等截面杆、载荷通过横截面形心、材料是均质的(即杆件的所有部分相同)。

(1) 该句为主从复合句，主句的结构是"主句+系动词+表语"，表语是由第一个 that 引导的表语从句 that the deformation of the bar be uniform，表语从句中的谓语使用虚拟语气。

(2) which in turn ... be homogeneous (that is, the same throughout all parts of the bar)是非限制性定语从句，修饰主句。

(3) 在 which 引导的非限制性定语从句中，that 引导了三个并列的宾语从句，作动词 requires 的宾语，三个宾语从句的谓语都采用了虚拟语气。

Questions

1. How can the fundamental concepts of stress and strain be illustrated?
2. What is called the stress and is commonly denoted by the Greek letter σ?
3. What are the definitions of normal stress and strain?
4. In what way can the internal stresses produced by the axial forces be investigated?

Unit 3 Steels

Without alloy irons and steels, the state of technology would be set back considerably [1]. Many varieties of alloys have been developed to meet specific needs of an advancing civilization. However, the availability of many varieties has often resulted in poor selection and excess cost for an unnecessary and expensive alloy material [2]. It is the responsibility of the designer and manufacturing engineer to be knowledgeable in this area and to make the best selection from the available alternatives [3].

Plain-Carbon Steel

Steel theoretically is an alloy of iron and carbon. When produced commercially, however, certain other elements-notably manganese, phosphorus, sulfur and silicon are present in small quantities. When these four foreign elements are present in their normal percentages, the product is referred to as plain-carbon steel. Its strength is primarily a function of its carbon content. Unfortunately, the ductility decreases as the carbon content is increased, and its hardenability is quite low. In addition, the properties of ordinary carbon steels are impaired by both high and low temperatures, and they are subject to corrosion in most environments.

Plain-carbon steels are generally classed into three subgroups, based on carbon content. Low-carbon steels have less than 0.30 percent carbon, possess good formability and weldability, but not enough hardenability to be hardened to any significant depth. Their structures usually are ferrite and pearlite, and the material generally is used as it comes from the hot-forming or cold-forming process. Medium-carbon steels have between 0.25 and 0.60 per cent carbon, and they can be quenched to form martensite or bainite if section size is small and a severe water or brine quench is used. The best balance of properties is attained at these carbon levels, the high fatigue and toughness of the low-carbon material being in good compromise with the strength and hardness that comes with higher carbon content [4]. These steels find numerous applications. High-carbon steels have more than 0.60 percent carbon; toughness and formability are quite low, but hardness and wear resistance are high. Severe quenches can form martensite, but hardenability is still poor.

Quench cracking is often a problem when the material is pushed to its limit. Plain carbon steels are the lowest-cost steel material and should be considered for many applications. Often, however, their limitations become restrictive [5]. When improved material is required, steels can be upgraded by the addition of one more alloying elements [6].

Alloy Steels

The differentiation between "plain carbon" and "alloy" steel is often somewhat arbitrary. Both contain carbon, manganese, and usually silicon. Copper and boron also are possible additions to both classes. Steels containing more than 1.65 percent manganese, 0.60 percent silicon, or 0.60 percent copper are designated as alloy steels. Also, steel is considered to be an alloy steel if a definite amount or minimum of other alloying element is specified or required. The most common alloy elements are chromium, nickel, molybdenum, vanadium, tungsten, cobalt, boron, and copper, as well as manganese, silicon, phosphorus, and sulfur in amounts greater than normally are present.

◇ *New Words and Expressions*

excess [ɪk'ses]	n. 过度，剩余；adj. 过度的，额外的
plain [pleɪn]	n. 平原，草原；adj. 简单的，普通的，朴素的
alternative [ɔːl'tɜːnətɪv]	n. 二中择一，可供选择的办法；adj. 备选的
manganese ['mæŋgəniːz]	n. [化] 锰(元素符号为 Mn)
phosphorus ['fɒsfərəs]	n. 磷
sulfur ['sʌlfə]	n. [化] 硫黄
silicon ['sɪlɪkən]	n. [化] 硅，硅元素
elements ['elɪmənts]	n. 原理，基础，元素
hardenability [hɑːdənə'bɪlɪtɪ]	n. 可硬化性，淬硬性
impair [ɪm'peə(r)]	v. 削弱
subgroup ['sʌbgruːp]	n. 小群，隶属的小组织，子群
martensite ['mɑːtənˌzaɪt]	n. [冶] 马氏体
bainite ['beɪnaɪt]	n. [冶] 贝菌体，贝氏体
brine [braɪn]	n. 盐水
compromise ['kɒmprəmaɪz]	n. & v. 妥协，折中，综合考虑，妥善
restrictive [rɪ'strɪktɪv]	adj. 限制性的
arbitrary ['ɑːbɪtrəri]	adj. 任意的，武断的，独裁的
chromium ['krəʊmiəm]	n. 铬
molybdenum [mə'lɪbdənəm]	n. [矿] 钼矿
cobalt ['kəʊbɔːlt]	n. [化] 钴
universal [juːnɪ'vɜːsl]	adj. 普遍的，全体的，通用的，宇宙的
identify [aɪ'dentɪfaɪ]	vt. 识别，鉴别；v. 确定
prefix ['priːfɪks]	n. [语] 前缀

⇨ *Notes*

[1] Without alloy irons and steels, the state of technology would be set back considerably.

参考译文：如果没有钢铁合金，技术的状态就会后退很多。

此句中 Without…表示假设条件，谓语为虚拟语气。

[2] However, the availability of many varieties has often resulted in poor selection and excess cost for an unnecessary and expensive alloy material.

参考译文：然而，材料品种多却常常导致不宜选材，或是选择不必要的昂贵合金材料而大大增加成本。

此句中介词 for 短语为原因状语。

[3] It is the responsibility of the designer and manufacturing engineer to be knowledgeable in this area and to make the best selection from the available alternatives.

参考译文：具备这方面的知识，并在现有钢材中选择最佳材料是设计工程师和制造工程师的职责。

此句中，It 是形式主语, to be knowledgeable…and to make … 是真实主语。

[4] The best balance of properties is attained at these carbon levels, the high fatigue and toughness of the low-carbon material being in good compromise with the strength and hardness that comes with higher carbon content.

参考译文：具有这样的碳含量的钢，能具备最好的综合性能，它既具有低碳钢的高疲劳强度和韧性，又兼备高碳钢的强度和硬度。

the high fatigue … being … 是 the best balance of properties 的同位语，对主语进一步补充说明。

[5] Often, however, their limitations become restrictive.

参考译文：然而，它们的局限性往往使它们的应用受到限制。

考虑到上一句的意思，翻译时加"使它们的应用"才能使意思更为明显。

[6] When improved material is required, steels can be upgraded by the addition of one more alloying elements.

参考译文：当需要改进这些材料时，可以加入一种或多种合金元素来提高钢的等级。

by the addition of…这一短语用作方式状语。

⊠ *Questions*

1. What is plain-carbon steel?
2. In what subgroups are plain-carbon steels classed into?
3. What is the difference between plain carbon steel and alloy steel?
4. What are the most common alloy elements of steel?

Reading Material: Selection of Construction Materials

There is not a great difference between "this" steel and "that" steel; all are very similar in mechanical properties. Selection must be made on factors such as hardenability, price, and availability, and not with the idea that "this" steel can do something no other can do because it contains 2 percent instead of 1 percent of a certain alloying element, or because it has a mysterious name. A tremendous range of properties is available in any steel after heat treatment; this is particularly true of alloy steels.

Considerations in Fabrication

The properties of the final part (hardness, strength, and machinability) rather than properties required by forging, govern the selection of material. The properties required for forging have very little relation to the final properties of the material; therefore, not much can be done to improve its forgeability. Higher-carbon steel is difficult to forge. Large grain size is best if subsequent heat treatment will refine the grain size.

Low-carbon, nickel-chromium steels are just about as plastic at high temperature under a single 520ft·lb blow as plain steels of similar carbon content. Nickel decreases formability of medium-carbon steels, but has little effect on low-carbon steels. Chromium seems to harden steel at forging temperatures, but vanadium has no discernible effect; neither has the method of manufacture any effect on high-carbon steel.

Formability

The cold-formability of steel is a function of its tensile strength combined with ductility. The tensile strength and yield point must not be high or too much work will be required in bending; likewise, the steel must have sufficient ductility to flow to the required shape without cracking. The force required depends on the yield point, because deformation starts in the plastic range above the yield point of steel. Work-hardening also occurs here, progressively stiffening the metal and causing difficulty, particularly in the low-carbon steels.

It is quite interesting in this connection to discover that deep draws can sometimes be made in one rapid operation that could not possibly be done leisurely in two or three[1]. If a draw is half made and then stopped, it may be necessary to anneal before proceeding, that is, if the piece is given time to work-harden. This may not be a scientific statement, but it is actually what seems to happen.

Internal Stresses

Cold forming is done above the yield point in the work-hardening range, so internal stresses can be built up easily. Evidence of this is the springback as the work leaves the forming operation and the warpage in any subsequent heat treatment. Even a simple washer might, by virtue of the internal stresses resulting from punching and then flattening, warp severely during heat treating. When doubt exists as to whether internal stresses will cause warpage, a piece can be checked by heating it to about 1,100°F and then letting it cool. If there are internal stresses, the piece is likely to deform. Pieces that will warp severely while being heated have been seen, yet the heat-treater was expected to put them through and bring them out better than they were in the first place[2].

Welding

The maximum carbon content of plain carbon steel safe for welding without preheating or subsequent heat treatment is 0.30 %. Higher-carbon steel is welded every day, but only with proper preheating. There are two important factors: the amount of heats that is put in; the rate at which it is removed.

Welding at a slower rate puts in more heat and heats a large volume of metal, so the cooling rate due to loss of heat to the base metal is decreased. A preheat will do the same thing. For example, sae 4150 steel, preheated to 600°F or 800°F, can be welded readily. When the flame or arc is taken away from the weld, the cooling rate is not so great, owing to the higher temperature of the surrounding metal and slower cooling results. Even the most rapid air-hardening steels are weldable if preheated and welded at a slow rate.

Machinability

Machinability means several things. To production men it generally means being able to remove metal at the fastest rate, leave the best possible finish, and obtain the longest possible tool life [3]. Machinability applies to the tool-work combination.

It is not determined by hardness alone, but by the toughness, microstructure, chemical composition, and tendency of a metal to harden under cold work. In the misleading expression "too hard to machine", the word "hard" is usually meant to be synonymous with "difficult". Many times a material is actually too soft to machine readily. Softness and toughness may cause the metal to tear and flow ahead of the cutting tool rather than cut cleanly. Metals that are inherently soft and tough are sometimes alloyed to improve their machinability at some sacrifice in ductility. Examples are use of lead in brass and of sulfur in steel.

Machinability is a term used to indicate the relative ease with which a material can be machined by sharp cutting tools in operations such as turning, drilling, milling, broaching, and reaming.

In the machining of metals, the metal being cut, the cutting tool, the coolant, the process and type of machine tool, and the cutting conditions all influence the results. By changing any one of these factors, different results will be obtained. The criterion upon which the ratings listed are based is the relative volume of various materials that may be removed by turning under fixed conditions to produce an arbitrary fixed amount of tool wear.

◇ *New Words and Expressions*

availability [əˌveɪlə'bɪlətɪ]   n. 可用性，有效性，可得性
fabrication [ˌfæbrɪ'keɪʃn]   n. 制造
forgeability [fɔ:dʒə'bɪlətɪ]   n. 可锻性
nickel ['nɪkl]   n. 镍
chromium ['krəʊmiəm]   n. 铬
vanadium [və'neɪdiəm]   n. 钒
discernible [dɪ'sɜ:nəbəl]   adj. 可辨别得出的，可看出的
cracking ['krækɪŋ]   n. 开裂，裂纹，裂缝
work-harden   加工硬化，冷作硬化
anneal [ə'ni:l]   n. & v. 退火
warp [wɔ:p]   v. 弯曲，变形
preheat [ˌpri:'hi:t]   v. 预热
microstructure ['maɪkrəʊˌstrʌktʃə]   n. 显微结构
mislead [ˌmɪs'li:d]   vt. 使……误解
ream [ri:m]   vt. (用铰刀)铰孔
arbitrary ['ɑ:bɪtrərɪ]   adj. 任意的

⇨ *Notes*

[1] It is quite interesting in this connection to discover that deep draws can sometimes be made in one rapid operation that could not possibly be done leisurely in two or three.

参考译文：在这方面，相当有趣的是你将发现有时可通过一次快速加载完成深度拉伸，但以缓慢的方式两、三次加载却不能实现。

in this connection 意为在这方面。deep draws 意为深度拉伸。

[2] Pieces that will warp severely while being heated have been seen, yet the heat-treater was expected to put them through and bring them out better than they were in the first place.

参考译文：即使是一个简单的垫圈，由于打孔和随后的平整加工中产生内应力，也会在热处理中呈现严重的翘曲。

[3] Machinability means several things. To production men it generally means being able to remove metal at the fastest rate, leave the best possible finish, and obtain the longest possible tool life.

参考译文：对于(机械加工)工人来说，可加工性通常意味着能够以最快的速度切削工件，获得最好的表面光洁度，并使刀具保持最长的使用寿命。

Questions

1. What factors should be considered in the selection of construction materials?
2. Why should fabrication be considered when selecting materials?
3. What does machinability mean?
4. What is cold forming?
5. What is the purpose of welding?

Unit 4 Cast Iron

Cast iron, essentially an alloy of iron, carbon, and silicon, is composed of iron and from 2 to 6.67 percent carbon, plus manganese, sulfur, and phosphorus. Commerecial cast iron contains no more than 4 percent carbon. Cast iron is often alloyed with elements such as nickel, chromium, molybdenum, vanadium, copper, and titanium. Alloying elements toughen and strengthen cast irons.

Gray Cast Iron

Gray cast iron is a relatively brittle material, mainly because of its long thin graphite flakes that are very weak. Gray cast iron is a metal that will withstand large compressive loads but small tensile loads.

White Cast Iron

White cast iron is very hard, brittle, and virtually nonmachinable. In some cases it is used where there is a need for resistance to abrasion. White cast iron is often found in combination with other cast iron, such as gray cast iron, to improve the hardness and wear resistant properties.

There are basically two ways of obtaining white cast iron. One way is by lowering the iron's silicon content; the second is by rapid cooling, which in this case yields what is called chilled cast iron. When cooled at a rapid rate, the excess carbon forms iron carbide and not graphite, thus making white cast iron.

Malleable Cast Iron

Malleable cast iron is noted for its strength, toughness, ductility, and machinability. In the process of making malleable cast iron, it is necessary to begin with white cast iron. The white cast iron is then heat treated as follows.

1. Heat to about 1700°F (927°C);

2. Hold at this temperature for about 15 hours. This breaks down the iron carbide to austenite and graphite;

3. Slow cool to about 1300°F (704°C) ;

4. Hold at this temperature for approximately 15 hours;

5. Air cool to room temperature.

The above process breaks down the iron carbide into additional austenite and graphite. Upon cooling the graphite will form into clusters or balls. The austenite will take on any one of the transformation products, depending on the cooling rate.

Nodular Cast Iron

Nodular cast iron is known by several names: nodular iron, ductile iron, and spheroidal graphite iron. It gets the names from the ball-like form of the graphite in the metal and the very ductile property it exhibits. Nodular cast iron combines many of the advantage of cast iron and steel. Its advantages include good castability, toughness, machinability, good wear resistance, weldability, low melting point, and hardenability.

The formation of the graphite into a ball form is accomplished by adding certain elements such as magnesium and cerium to the melt just prior to casting. The vigorous mixing reaction caused by adding these elements results in a homogeneous spheroidal or ball-like structure of the graphite in the cast iron. The iron matrix or background material can be heat treated to form any one of the microstructures associated with steels such as ferrite, pearlite, or martensite.

◇ New Words and Expressions

titanium [taɪˈteɪnɪəm]	n.	钛(Ti)
alloy [ˈælɒɪ]	v.	合铸，熔合，加合金元素
toughen [ˈtʌfn]	v.	韧化，使坚韧
graphite [ˈɡræfaɪt]	n.	石墨
flake [fleɪk]	n.	薄片
abrasion [əˈbreɪʒ(ə)n]	n.	磨损
chill [tʃɪl]	v.	激冷
malleable [ˈmælɪəb(ə)l]	adj.	可锻的，有韧性的
machinability [məˌʃiːnəˈbɪlətɪ]	n.	可切削性，切削加工性
cluster [ˈklʌstə]	n.	线束，簇
nodular [ˈnɔdjulə, -dʒə-]	adj.	球状的
spheroidal [sfɪəˈrɒɪdəl]	adj.	(扁)球体的，球状的
castability [ˌkaːstəˈbɪləti]	n.	可铸性
weldability [weldəˈbɪlɪtɪ]	n.	可焊性，焊接性
magnesium [mæɡˈniːzɪəm]	n.	镁(Mg)
cerium [ˈsɪərɪəm]	n.	铈(Ce)
vigorous [ˈvɪɡ(ə)rəs]	adj.	活泼的
prior to		在……之前
matrix [ˈmeɪtrɪks]	n.	基体
(be) associated with		与……有关

✕ Questions

1. What advantages does nodular cast iron have?
2. What is the definition of gray cast iron?
3. What other names does nodular cast iron have?
4. In what way is the white cast iron heat treated?
5. What features does white cast iron have?

Reading Material: Metals and Their Use

It is known that metals are very important in our life. Metals have the greatest importance for industry. All machines and other engineering constructions have metal parts, some of them consist only of metal parts.

There are two large groups of metals;
1) Simple metals—more or less pure chemical elements.
2) Alloys—materials consisting of a simple metal combined with some other elements.

About two thirds of all elements found in the earth are metals, but not all metals may be used in industry [1]. Those metals which are used in industry are called engineering metals. The most important engineering metal is iron (Fe), which in the form of alloys with carbon(C) and other elements, finds greater use than any other metal. Metals consisting of iron combined with some other elements are known as ferrous metals [2]; all the other metals are called nonferrous metals. The most important nonferrous metals are copper(Cu), aluminum(Al), lead(Pb), zinc(Zn), tin(Sn), but all these metals are used much less than ferrous metals, because the ferrous metals are much cheaper.

Engineering metals are used in industry in the form of alloys because the properties of alloys are much better than the properties of pure metals. Only aluminum may be largely used in the form of a simple metal. Metals have such a great importance because of their useful properties or their strength, hardness, and plasticity.

Different metals are produced in different ways, but almost all the metal are found in the form of metal ore (iron ore, copper ore, etc.).

The ore is a mineral consisting of a metal combined with some impurities. In order to produce a metal from some metal ore, we must separate these impurities from the metal, which is done by metallurgy.

◇ New Words and Expressions

zinc [zɪŋk]   n. 锌
construction [kən'strʌkʃ(ə)n] n. 结构
tin [tɪn]   n. 锡

metallurgical [ˌmetə'lɜːdʒɪkl]	adj. 冶金(学)的
iron ['aɪən]	n. 铁
strength [streŋθ; streŋkθ]	n. 强度，实力
carbon ['kɑːb(ə)n]	n. 碳
ferrous ['ferəs]	adj. (含)铁的，二价铁的
plasticity [plæ'stɪsɪtɪ]	n. 可塑(适应)性
ore [ɔː]	n. 矿，矿砂(石)
copper ['kɒpə]	n. 铜
impurity [ɪm'pjʊərɪtɪ]	n. 杂质，不纯
aluminum [ə'luːmɪnəm]	n. 铝
mineral ['mɪn(ə)r(ə)l]	n. 矿物
lead [liːd]	n. 铅
combine...with	把……和……结合(起来)
nonferrous metals	有色金属

⇨ Notes

[1]　About two thirds of all elements found in the earth are metals, but not all metals may be used in industry.

参考译文：地球上发现的元素中大约三分之二是金属元素，但不是所有的金属都能够用于工业上。

found in the earth 是过去分词短语，作 all elements 的后置定语，two thirds 意为"三分之二"。

[2]　Metals consisting of iron combined with some other elements are known as ferrous metals.

参考译文：由铁跟某种其它元素相结合组成的金属称为黑色金属。

consisting of 是短语动词的现在分词，它与后面的宾语组成现在分词短语作 metals 的后置定语。本句的短语动词是 are known as。

⊠ Questions

1. How many groups of metals are there?
2. Why are metals very important in our life?
3. What is the most important engineering metal?
4. Why do metals have such a great importance in engineering?
5. What is the difference between simple metals and alloys?

Unit 5 Stainless Steels

Stainless steels do not rust in the atmosphere as most other steels do. The term "stainless" implies a resistance to staining, rusting, and pitting in the air, moist and polluted as it is, and generally defines a chromium content in excess of 11% but less than 30%. And the fact that the stuff is "steel" means that the base is iron.

Stainless steels have room-temperature yield strengths that range from 205 MPa (30 ksi) to more than 1725 MPa (250 ksi). Operating temperatures around 750℃ (1400 °F) are common, and in some applications temperatures as high as 1090℃(2000°F) are reached. At the other extreme of temperature some stainless steels maintain their toughness down to temperatures approaching absolute zero.

With specific restrictions in certain types, the stainless steels can be shaped and fabricated in conventional ways. They can be produced and used in the as-cast condition; shapes can be produced by powder-metallurgy techniques; cast ingots can be rolled or forged (and this accounts for the greatest tonnage by far). The rolled product can be drawn, bent, extruded, or spun. Stainless steel can be further shaped by machining, and it can be joined by soldering, brazing, and welding. It can be used as an integral cladding on plain carbon or low alloy steels.

The generic term "stainless steel" covers scores of standard compositions as well as variations bearing company trade names and special alloys made for particular applications. Stainless steels vary in their composition from a fairly simple alloy of, essentially, iron with 11% chromium, to complex alloys that include 30% chromium, substantial quantities of nickel and half a dozen other effective elements. At the high-chromium, high-nickel end of the range they merge into other groups of heat-resisting alloys, and one has to be arbitrary about a cutoff point. If the alloy content is so high that the iron content is about half, however, the alloy falls outside the stainless family. Even with these imposed restrictions on composition, the range is great, and naturally, the properties that affect fabrication and use vary enormously. It is obviously not enough to specify simply a "stainless steel".

The various specifying bodies categorize stainless steels according to chemical composition and other properties. For example, the American Iron and Steel Institute (AISI) lists more than 40 approved wrought stainless steel compositions; the American Society for Testing and Materials (ASTM) calls for specifications that may conform to AISI compositions but additionally require certain mechanical properties and dimensional tolerances; the Alloy Casting Institute (ACI) specifies compositions for cast stainless steels within the categories of corrosion-and heat-resisting alloys; the Society of Automotive Engineers (SAE) has adopted AISI and ACI compositional specifications. Military specification MIL-HDBK-5 lists design values. In addition, manufacturers' specifications are used for special purposes or for proprietary alloys. Federal and military specifications and manufacturers' specifications are laid down for special purposes and sometimes acquire a general acceptance.

However, all the stainless steels, whatever specifications they conform to, can be conveniently classified into six major classes that represent three distinct types of alloy constitution, or structure. These classes are ferritic, martensitic, austenitic, manganese-substituted austenitic, duplex austenitic-ferritic, and precipitation-hardening.

Ferritic Stainless steel is so named because the crystal structure of the steel is the same as that of iron at room temperature. The alloys in the class are magnetic at room temperature and up to their Curie temperature (about 750℃; 1400 ºF). Common alloys in the ferritic class contain between 11% and 29% chromium, no nickel, and very little carbon in the wrought condition. The 11% ferritic chromium steels, which provide fair corrosion resistance and good fabrication at low cost, have gained wide acceptance in automotive exhaust systems, containers, and other functional applications. The intermediate chromium alloys, with 16%～17% chromium, are used primarily as automotive trim and cooking utensils, always in light gages, their use somewhat restricted by welding problems. The high-chromium steels, with 18%～29% chromium content, have been used increasingly in applications requiring a high resistance to oxidation and, especially, to corrosion. These alloys contain either aluminum or molybdenum and have low carbon content.

The high-temperature form of iron (between 910℃ and 1400℃, or 1670 ºF and 2550ºF) is known as austenite (Strictly speaking the term austenite also implies carbon in solid solution). The structure is nonmagnetic and can be retained at room temperature by appropriate alloying. The most common austenite retainer is nickel. Hence, the traditional and familiar austenitic stainless steels have a composition that contains sufficient chromium to offer corrosion resistance, together with nickel to ensure austenite at room temperature and below. The basic austenitic composition is the familiar 18% chromium, 8% nickel alloy. Both chromium and nickel contents can be increased to improve corrosion resistance, and additional elements (most commonly molybdenum) can be added to further enhance corrosion resistance.

◇ *New Words and Expressions*

as-cast ['æsk'ɑːst]	adj. 铸态的
powder-metallurgy	n. 粉末冶金学
cast ingot [kɑːst'ɪŋgət]	n. 铸锭
roll [rəul]	n. 轧制
tonnage ['tʌnɪdʒ]	n. (总)吨位
extrude [ɪk'struːd]	vt. 挤压

spin [spɪn]	v. 旋压
solder ['səʊldə(r)]	vt. 钎焊
braze [breɪz]	vt. 铜焊
cladding ['klædɪŋ]	n. 包层，覆盖，(金属)覆层
wrought [rɔːt]	adj. 可锻的
American Iron and Steel Institute (AISI)	美国钢铁学会
American Society for Testing and Materials(ASTM)	美国材料试验学会
Alloy Casting Institute (ACI)	合金铸造学会
Society of Automotive Engineers (SAE)	美国汽车工程师学会
ferritic [fə'rɪtɪk]	adj. 铁素体的
martensitic [mɑːtɪn'zɪtɪk]	adj. 马氏体的
austenitic [ˌɔːstə'nɪtɪk]	adj. 奥氏体的
oxidation [ˌɒksɪ'deɪʃn]	n. 氧化

Questions

1. What special features do stainless steels have?
2. What is ferritic stainless steel?
3. What are the obvious limitations of stainless steels?

Reading Material: Selection of Stainless Steel

The justification for selecting stainless steel is corrosion and oxidation resistance. Stainless steels possess, however, other outstanding properties that in combination with corrosion resistance contribute to their selection. These are the ability to develop very high strength through heat treatment or cold working, weldability, formability and in the case of austenitic steels, low magnetic permeability and outstanding cryogenic mechanical properties.

The choice of a material is not simply based on a single requirement, however, even though a specific condition (for example, corrosion service) may narrow the range of possibilities. For instance, in the choice of stainless steel for railroad cars, while corrosion resistance is one determining factor, strength is particularly significant. The higher price of stainless steel compared with plain carbon steel is moderated by the fact that the stainless has about twice the allowable design strength. This not only cuts the amount of steel purchased, but by reducing the dead weight of the vehicle, raises the load that can be hauled. The same sort of reasoning is even more critical in aircraft and space vehicles.

But weight saving alone may be accomplished by other materials, for example, the high-strength low-alloy steels in rolling stock and titanium alloys in aircraft. Thus, the selection of a material involves a careful appraisal of all service requirements as well as a consideration of the ways in which the required parts can be made. It would be foolish to select material on the basis of its predicted performance if the required shape could be produced only with such difficulty that cost skyrocketed.

The applicability of stainless steels may be limited by some specific factors, for example, an embrittlement problem or susceptibility to a particular corrosive environment. In general terms, the obvious limitations are:

① In chloride environments susceptibility to pitting or stress-corrosion cracking requires carful appraisal. One cannot blindly assume that a stainless steel of some sort will do. In fact, it is possible that no stainless steel will serve.

② The temperature of satisfactory operation depends on the load to be supported, the time of its application, and the atmosphere. However, to offer a round number for the sake of marking a limit, we suggest a maximum temperature of 870℃ (1600℉). Common stainless steels can be used for short times above this temperature, or for extended periods if the load is only a few thousand pounds per square inch. But if the loads or the operating periods are great, then more exotic alloys are called for.

◆ New Words and Expressions

chloride ['klɔːraɪd]	n. 氯化物
cryogenic [ˌkraɪə'dʒenɪk]	adj. 低温的，深冷的
justification [ˌdʒʌstɪfɪ'keɪʃn]	n. 辩解，无过失，正当的理由
susceptibility [səˌseptə'bɪləti]	n. 易受影响或损害的状态，感受性

☒ Questions

1. Can we select material on the basis of its predicted performance?
2. How can all the stainless steels be conveniently classified?
3. How can weight saving be accomplished?

PART 2   Forming Technology

Unit 6 Casting

Metal casting is one of the oldest of all industries, both ancient and medieval history offering examples of the manufacture and use of casting. From simple axeheads poured from copper in open moulds some 5000 years ago, casting in the pre-Christian world developed to a point at which elaborate bronze statuary could be produced in two-piece and cored moulds. By the end of the medieval period, decorated bronze and pewter castings had begun to be used in European church and domestic life.

The widespread adoption of cast iron as engineering material awaited the success of Abraham Darby in 1709 in smelting in the coke blast furnace; this paved the way for the massive use of cast iron in construction during the years following the industrial revolution.

Many foundries sprang up after the industrial revolution, the vast majority being for the manufacture of the cast iron then being used as a structural material. The quantity production of iron castings in the nineteenth century was not matched by a universal advance in quality and the engineering use of the products encountered more serious risks in a non-ductile material.

Despite the skill of the molder in producing complex forms, there was little change in the metallurgical and engineering situation until the modern era brought a better understanding of the factors determining quality. With modern techniques of process control the rudimentary judgment of the operator could give way to objective measurements of metal temperature, molding material properties and other production variables. These improvements have been applied not only to cast iron but to a wide range of cast alloys.

There are four basic casting methods: sand-casting, die-casting, investment-casting, and centrifugal casting.

Sand-casting is the most widely used method employed in foundry. In this process, sand moulds are contained in metal molding boxes that have four sides but top or bottom. During the molding operation the boxes are located together by pins so that they can be separated to remove the pattern, and replaced in the correct position before the metal is poured in. The boxes are clamped together, or the cope (top section) weighted down when pouring to prevent the cope from "floating away" from the drag (lower section) when the mould is full of molten metal. The sequence when molding the simple two-part mould to cast a bracket is illustrated as follows.

At the first stage the pattern is seated on the moulding board. The pattern is covered with facing sand, which is a specially prepared sand of good quality, which can take a clean and smooth impression, and can resist the heat from the metal that will be in contact with it. The facing sand is backed up with molding sand, which is old facing sand from previous moulds. The molding sand is carefully rammed up so that it is fairly tight around the pattern to produce a good solid mould, yet permeable enough to allow the gases produced during casting to escape. The sand is finally leveled off.

At the second stage the mould with the pattern still in position is inverted; the exposed sand lightly covered with parting sand, and the exposed pattern with facing sand. (The parting sand has no cohesion, and is introduced to permit a clean separation when the mould is opened up to remove the pattern.) The second molding box is located in position on the first box and filled with molding sand. Two or more plugs are introduced when the second box is being filled (these are removed later, leaving channels in the sand). One of these plugs is positioned to one side of the pattern. The sand is rammed up and leveled off.

Now, at stage 3, allow the pattern to be removed. This is done by screwing a bar with a threaded end into a suitable insert in the pattern, damping the sand around the pattern, and gently rapping the bar in all directions so that the pattern can be carefully withdrawn. To facilitate the removal of the pattern without scuffing the sides of the impression, all surfaces that lie in the direction of pattern removal are inclined slightly by a small amount (the draw angle).

A groove called a gate is cut in the sand face to allow the channel produced by the plug that is outside the pattern to connect with the impression. The metal is poured through this channel (called the runner), and the gate prevents it from dropping straight into the impression and damaging it. The cross-section of the gate is slightly smaller than that of channel so that a full runner will always supply metal to the gate at a slight pressure.

Finally, the mould is reassembled, carefully locating and securing the two sections. The top section is known as the cope, and the lower section is known as the drag. The sand in the cope is vented. These vents allow the sand to be rammed up more tightly at the earlier stages without the risk of gases being trapped in the molten metal and forming blowholes in the solid metal. A sand-feeding gate (also called a pouring or bowl) is added to make it easier to pour the metal into the runner. The molten metal is poured through the runner and the air will escape through the riser.

The impression will be filled with molten metal when it is completely filled. Gases can escape through the runner and the riser, which also act as headers to supply the impression with more metal to compensate for the contraction of the metal when cooling in the molten state.

◇ *New Words and Expressions*

    casting ['ka:stiŋ]   n. 铸造 (件)
    medieval [ˌmedi'i:vəl]   adj. 中世纪的，老式的
    elaborate [i'læbərət]   adj. 精心制作的，精细的
    bronze [brɔnz]   n. 青铜 (铜与锡的合金)
    pewter ['pju:tə]   n. 白镴，锡铅合金器皿
    metallurgical [ˌmetə'lə:dʒikəl]   adj. 冶金学的
    rudimentary [ru:di'mentəri]   adj. 根本的，未发展的

permeable ['pə:miəbl]	adj. 有浸透性的，能透过的
cohesion [kəu'hi:ʒən]	n. 结合，凝聚
scuffing ['skʌfiŋ]	n. 刮(磨，擦，划)伤
runner ['rʌnə(r)]	n. 浇口
riser ['raizə]	n. 冒口
header ['hedə]	n. 内浇口
	炼焦炉
sand-cast	砂模铸造，翻砂
die-casting	金属模铸造
	熔模铸造
clean separation	清晰分离，完全分离

❯ Questions

1. How many casting methods are there? What are they?
2. Please describe the history of metal casting.
3. Why is sand-casting the most widely used method in foundry?
4. What is the sequence of molding two-part mould to cast a bracket?
5. What is sand-casting?

Reading Material: Ingot Casting

In general, most steel is produced by the basic-oxygen or open hearth processes and is transferred in the molten state in large ladles. The steel is then cast or "teemed" into large stationary ingot molds, or cast continuously into long ingots.

Individual Ingot Casting

Casting the steel into individual ingot molds is the conventional method of producing steel ingots for hot working, and most steel is cast in this way since the method is so versatile. In this process, the full ladle of steel is moved by overheat crane so that it can be tapped (or teemed) into individual molds standing upright on rail cars. The ingot molds are slightly tapered for easy removal after solidification of the steel. After stripping the ingot molds, the hot ingots are transformed to soaking pits for hot rolling. About 85 percent of the steel cast today is still cast in individual molds.

Continuous Casting

In continuous casting, the ladle of molten steel is transported to an elevated casting platform above a casting machine. The molten steel is discharged into a rectangular trough, called a tundish, which acts as a reservoir for the steel. Form a spout in the bottom of the tundish, the molten steel is slowly lowered. As the molten steel enters the mold, the metal at the mold surface solidifies, forming a thin skin. This skin thickens as the metal passes through the mold. The remaining metal in the center of the ingot is solidified by cold water sprayed onto the ingot as it leaves the mold.

The solid metal billet is pulled by rollers so that a long continuous steel slab is produced. For many types of steel, this process is more economical than stationary casting into individual molds. More steel can be cast into slab form in a shorter time than with the individual molds. The metal does not have to be steel industry today is to use continuous casting wherever possible since it can in many cases produce a higher quality product at lower cost .

◇ *New Words and Expressions*

ladle ['leɪdl]	n. 长柄勺
teem [ti:m]	v. 把……注入模具
versatile ['vɜ:sətaɪl]	adj. (指工具、机器等)多用途的
crane [kreɪn]	n. 吊车，起重机
taper ['teɪpə(r)]	vt. & vi. 逐渐变细，变尖；逐渐减弱
strip [strɪp]	vt. 除去，剥去；清除，拆除
discharge [dɪs'tʃɑ:dʒ]	vt. 放出；流出
trough [trɒf]	n. 水槽，食槽
tundish ['tʌndɪʃ]	n. 漏斗
reservoir ['rezəvwɑ:(r)]	n. 蓄水池；贮液器
spout [spaʊt]	n. 喷口，喷嘴
solidify [sə'lɪdɪfaɪ]	vt. 使凝固，固化；使结晶
billet ['bɪlɪt]	n. 钢胚

⇨ *Notes*

[1] Casting the steel into ingot molds is the conventional method of producing steel ingots for hot working......

参考译文：把钢锭浇铸成钢锭模是生产热锻钢锭的常规方法。

[2] In a continuous casting, the ladle of molten steel is transported to an elevated casting platform above a casting machine.

参考译文：在连续铸造中，钢水包被输送到铸造机上方的高架铸造平台上。

✕ *Questions*

1. What is the conventional method of producing steel ingots for hot working?

2. What is individual ingot casting?
3. Please explain the working principle of continuous casting.
4. Why is continuous casting more economical than stationary casting into individual molds?
5. Please describe the process of converter.

Unit 7 Forging

Press forging employs a slow squeezing action in deforming the plastic metal, as contrasted with the rapid-impact blows of a hammer [1]. The squeezing action is carried completely to the center of the part being pressed, thoroughly working the entire section. These presses are the vertical type and may be either mechanically or hydraulically operated. The mechanical presses, which are faster operating and most commonly used, range in capacity from 500 to10, 000 tons.

For small press forgings closed impression dies are used, and only one stroke of the ram is normally required to perform the forging operation. The maximum pressure is built up at the end of the stroke which forces the metal into shape. Dies may be mounted as separate units, or all the cavities may be put into a single block. For small forgings individual die units are more convenient. There is some difference in the design of dies for different metals; copper-alloy forgings can be made with less draft than steel, consequently more complicated shapes can be produced. These alloys flow well in the die and are rapidly extruded.

In the forging press a greater proportion of the total work put into the machine is transmitted to the metal than in a drop hammer press [2]. Much of the impact of the drop hammer is absorbed by the machine and foundation. Press reduction of the metal is faster, and the cost of operation is consequently lower. Most press forgings are symmetrical in shape, having surfaces which are quite smooth, and provide a closer tolerance than is obtained by a drop hammer. However, many parts of irregular and complicated shapes can be forged more economically by drop forging. Forging presses are often used for sizing operations on parts made by other forging processes.

In drop forging, a piece of metal, roughly or approximately of the desired shape, is placed between die faces having the exact form of the finished piece, and forced to take this form by drawing the dies together. This method is widely used for the manufacture of parts both of steel and brass. Large ingots are now almost always forged with hydraulic presses instead of with steam hammers, since the work done by a press goes deeper [3]. Further, the press can take a cooler ingot and can work to closer dimensions. The forging should be done at about the same temperature as tolling; the process improves the physical properties of the steel just as rolling does [4]. In the final forging it is important not to have the steel too hot, for an overheated steel will have poor mechanical properties when cooled [5]. In heating for forging the temperature is usually judged by the eye, but where large numbers of the same pattern are made, the piece to be forged is heated in furnaces in which the temperature is indicated by pyrometers, and often is automatically controlled.

◇ *New Words and Expressions*

press [pres]	n. 压力机；v. 压锻
forging ['fɔːdʒɪŋ]	n. & adj. 锻造(件)，锻造的
roughly ['rʌflɪ]	adv. 粗(糙，略)地，大体上
deform [dɪ'fɔːm]	vt. & vi. (使)变形
desire [dɪ'zaɪə]	v. &n. 愿(期，希)望，要(请)求
shape [ʃeɪp]	n. 形(状，态)，模(造)形型腔模具
exact [ɪg'zækt]	adj. 精(准，正)确的；vt. 强制(坚持)要求
stroke [strəʊk]	n. 冲程；vt. 冲击
ram [ræm]	n. 锤头，冲头
dies ['daɪz]	n. 模具，模子
cavity ['kævɪtɪ]	n. 腔，窝，中空
block [blɒk]	n. 坯料，毛坯
brass [brɑːs]	n. 黄铜，黄铜制品
indicate ['ɪndɪket]	vt. 指示，显示
hydraulic [haɪ'drɔlɪk]	adj. 液(水)压的，液压传动装置
draft [drɑːft]	n. 牵引(力，阻力)
close [kləʊs]	adj. 接近的，精细的
complicate ['kɒmplɪkeɪt]	v. (使)复杂化
rolling ['rəʊlɪŋ]	n. 压光，轧制
extrude [ɪk'strʊd]	v. 挤压(成形)，模(热)压
impact ['ɪmpækt]	n. 碰撞，冲击
overheat [əʊvə'hiːt]	v. & n. 过热
hammer ['hæmə]	n. 落锤，锤头
judge [dʒʌdʒ]	v. 评判，判断
symmetrical [sɪ'mɛtrɪkl]	adj. 对称的
furnace ['fɜnɪs]	n. (熔，炼，高)炉
irregular [ɪ'rɛgjələ]	adj. 不规则的，不对称的
pyrometer [paɪ'rɒmɪtə]	n. 高温计
drop forging	落锤锻
closed impression dies	封闭式(扣盒式)
the finished piece	成品
drawing the dies together	使两模合拢
forced to take this form	压力成型

by a press goes deeper	深度持续施压作用
instead of…	代替，而不(是)
can take…	能处理(拿下)
the final forging	终锻，精锻
sizing operations	精锻工艺，精加工

➪ Notes

[1] Press forging employs a slow squeezing action in deforming the plastic metal, as contrasted with the rapid-impact blows of a hammer.

参考译文：压力锻与落锤锻的锤头快速降落的冲击锻造相反，是应用缓慢的挤压作用来使塑性金属变形。

press forging…, as contrasted with …a hammer 中 as contrasted with 是短语介词，表示"对照"、"对比"、"相对立"之意，可译为"与……相反""和……大不相同"；此短语介词与后面紧跟着一个短语一起组成一个介词短语，意思是"锤头的快速降落的冲击"，和主句中的"压力锻应用缓慢挤压作用"大不相同，或者是正相反的作用。

[2] In the forging press… the total work put into the machine is transmitted to…press.

参考译文：锻压机比落锤锻输入到机器里的总能量中更大部分的能量被传输到金属坯料上。

put into the machine 是过去分词短语作 the total work 的后置定语，而全句的谓语动词是 is transmitted to 即"被传输(递)到……"。

[3] …, since the work done by a press goes deeper.

参考译文：因为压力机所加工的工件可锻造得更加深透。

Since the work done by a press goes deeper 是从属连词 since 引导的状语从句。从句中主语为 the work，而 done by a press 为过去分词短语作 the work 的后置定语；该句的谓语是 goes，译为"忍受""承受""得到"，而 deeper 是副词比较级作状语修饰 goes。

[4] The forging should be done at about the same temperature as tolling; the process improves the physical properties of the steel just as rolling does.

参考译文：锻造将在大约与辊轧同样的温度下进行，此加工过程正像轧制金属那样可改善金属的物理性能。

just as 中 as 是从属连词，它引导一个作"比较"用的状语句 rolling does；该从句是省略从句，原句为 rolling improves the physical properties。

[5] In the final forging it is important not to have the steel too hot, for an overheated steel will have poor mechanical properties when cooled.

参考译文：精锻时，不要使钢件太热，那是很重要的，因为一件过热的钢件，它冷却后，其力学性能较差。

In the final forging… when cooled 中的 not 是否定 to have，不是否定 is 的；not to have 可译为"不使……"；而 when cooled 是一个省略了的时间状语从句，原句为 when the steel is cooled。

❯ Questions

1. What is press forging?
2. What should be avoided in the final forging?
3. What is the advantage of press reduction of the metal?
4. What is the process of forging operation?

Reading Material: Forging

Forging is the plastic working of metal by means of localized compressive forces exerted by manual or power hammers, presses, or special forging machines. It may be done either hot or cold. However, when it is done cold, special names usually are given to the processes. Consequently, the term "forging" usually implies hot forging done above the recrystallization temperature.

Modern forging is a development from the ancient art practiced by the armor makers and the immortalized village blacksmith. High-powered hammers and mechanical presses have replaced the strong arm, the hammer, and the anvil and modern metallurgical knowledge supplements the art skill of the craftsman in controlling the heating and handling of the metal.

Various materials respond quite differently to effort to deform them. It is particularly important that the forging producer has information on how readily a metal or alloy will deform during forging without exhibiting adverse effects, since the characteristic behavior of the forging material influence the choice of forging method, selection of equipment and die design. It is also important for the purchaser of forgings to recognize that forge ability varies among materials and affects forging design and processing costs [1].

Forge ability is the term used in the industry to denote a material relative resistance to deformation and its plasticity. While considerable disagreement exists as to precisely what characteristics the word "forge ability" should include, the term as used here is defined as the tolerance of a metal or alloy for deformation without failure, regardless of forging pressure requirements.

Raw material used for forging is generally bar or billet stock hot rolled from ingots melted in open-hearth, electric arc, or vacuum arc furnace. Other forms and shapes such as rolled slabs, plats, and stock produced by continuous casting techniques are occasionally used. For certain grades, vacuum arc melting imparts better forge ability than conventional arc melting. However, the major purpose of vacuum melting is the improvement of mechanical properties and cleanliness, not forging behavior.

Equipment behavior influences the forging progress since it determines the feasibility of forging a part and affects the rate of deformation and the temperature conditions.

The hammer is the most economical type of equipment for generating load and energy necessary to carry out a forging process, provided that the material being forged can support high deformation velocities. It is most commonly used hot forging equipment for repeated blows on the same workpiece and cannot be overloaded.

There are various types of hammers: air-lift gravity drop hammers, power drop hammers, counterblow hammers etc. In a simple gravity drop hammer the upper ram is positively connected to a board, a belt, a chain or a piston. When forging the ram is lifted to a certain height and then dropped on the stock placed on the anvil. During the down stroke, the ram is accelerated by gravity and builds up the blow energy. The upstroke takes place immediately after the blow, the force necessary to ensure quick lift-up of the ram can be 3 to 5 times the ram weight. The operation principle of a power-drop hammer is similar to that of an air drop hammer. During the down stroke, in addition to gravity, the ram is accelerated by steam, cold air or hot air pressure. In an elector hydraulic gravity-drop hammer, the ram is lifted with oil pressure against an air cushion. The compressed air slows down the upstroke of the ram and contributes to its acceleration during the down stroke. Thus, the electro hydraulic hammer also has a minor power hammer action.

◇ New Words and Expressions

recrystallize [ri:'kristəlaiz] v. (使)再结晶
armor ['a:mə] n. 铠甲
immortalize [i'mɔtəlaiz] v. 使不朽(灭)
anvil ['ænvil] n. (锤)砧, 砧座
vertical ['və:tikəl] adj. 立式的
craftsman ['kra:ftsmən] n. 技工
handling ['hændliŋ] n. 处理, 装卸, 搬运
exhibit [ig'zibit] v. 展示
denote [di'nəut] vt. 指示, 表示
billet ['bilit] n. 坯段
ram [ræm] n. 锤头, 滑块, 活动横梁
feasibility [ˌfi:zə'biləti] n. 可行性, 可能性
symmetrical [si'metrikəl] adj. 轴对称的
block [blɔk] n. (模)块
cushion ['kuʃən] n. 垫子, 软垫, 衬垫
sizing ['saiziŋ] n. 整形, 校正, 定径
flexibility [ˌfleksə'biliti] n. 柔韧性, 灵活性
furnace ['fə:nis] n. (熔, 高)炉

pyrometer ['paiə'rɔmitə]	n. 高温计
impact ['impækt]	n. 碰撞，冲击
mechanical press [mi'kænikəl] [pres]	机床压力机
hydraulic press [hai'drɔ:lik] [pres]	液压机
drop forging	落锻，锤模锻
rapid impact blow	快速冲击，猛打
impression die forging	模锻
closed impression die	闭式锻模

⇨ Notes

[1] It is also important for the purchaser of forgings to recognize that forge ability varies among materials and affects forging design and processing costs.

参考译文：对锻件的采购员来说，重要的是要认识到不同的材料锻造能力不同，并影响锻造设计和加工成本。

句中 forge ability 指锻件在锻造时形状和特性变化的能力，译为"可锻性"。

⊠ Questions

1. What is the definition and implication of forging?
2. What does modern forging originate from?
3. What types of hammers are there?
4. What does forge ability mean?
5. What are features of forging process?

Unit 8 Heat Treatment of Metals

The understanding of heat treatment is embraced by the broader study of metallurgy. Metallurgy is the physics, chemistry, and engineering related to metals from ore extraction to the final product. Heat treatment is the operation of heating and cooling a metal in its solid state to change its physical properties. According to the procedure used, steel can be hardened to resist cutting action and abrasion, or it can be softened to permit machining. With the proper heat treatment internal stresses may be removed, grain size reduced, toughness increased, or a hard surface produced on a ductile interior. The analysis of the steel must be known because small percentages of certain elements, notably carbon, greatly affect the physical properties.

Alloy steel owes their properties to the presence of one or more elements other than carbon, namely nickel, chromium, manganese, molybdenum, tungsten, silicon, vanadium, and copper [1]. Because of their improved physical properties they are used commercially in many ways not possible with carbon steels.

The following discussion applies principally to the heat treatment of ordinary commercial steels known as plain carbon steels. With this process, the rate of cooling is the controlling factor, rapid cooling from above the critical range results in hard structure, whereas very slow cooling produces the opposite effect.

A Simplified Iron-carbon Diagram

If we focus only on the materials normally known as steels, a simplified diagram is often used. Those portions of the iron-carbon diagram near the delta region and those above 2% carbon content are of little importance to the engineer and are deleted [2]. A simplified diagram, such as the one in Figure 8-1, focuses on the eutectoid region and is quite useful in understanding the properties and processing of steel.

The key transition described in this diagram is the decomposition of single-phase austenite (γ) to the two-phase ferrite plus carbide structure as temperature drops. Control of this reaction, which arises due to the drastically different carbon solubility of austenite and ferrite, enables a wide range of properties to be achieved through heat treatment.

To begin to understand these processes, consider a steel of the eutectoid composition, 0.77% carbon, being slow cooled along line x-x' in Figure 8-1. At the upper temperatures, only austenite is present, the 0.77% carbon being dissolved in solid solution with the iron. When the steel cools to 727℃(1341℉), several changes occur simultaneously. The iron wants to change from the FCC austenite structure to the BCC ferrite structure, but the ferrite can only contain 0.02% carbon in solid solution[3]. The rejected carbon forms the carbon-rich cementite intermetallic with composition Fe₃C. In essence, the net reaction at the eutectoid is Austenite (0.77%C)→ferrite (0.02%C) +cementite (6.67%C).

Since this chemical separation of the carbon component occurs entirely in the solid state, the resulting structure is a fine mechanical mixture of ferrite and cementite. Specimens prepared by polishing and etching in a weak solution of nitric acid and alcohol reveal the lamellar structure of alternating plates that forms on slow cooling. This structure is composed of two distinct phases, but has its own set of characteristic properties and goes by the name pearlite, because of its resemblance to mother- of- pearl at low magnification [4].

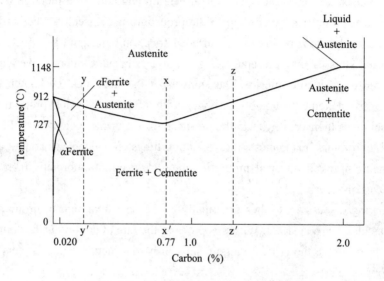

Figure 8-1　Simplified iron-carbon diagram

Steels having less than the eutectoid amount of carbon (less than 0.77%) are known as hypo-eutectoid steels. Consider now the transformation of such a material represented by cooling along line y-y' in Figure 8-1. At high temperatures, the material is entirely austenite, but upon cooling enters a region where the stable phases are ferrite and austenite. Tie-line and level-law calculations show that low-carbon ferrite nucleates and grows, leaving the remaining austenite richer in carbon [5]. At 727℃ (1341℉), the austenite is of eutectoid composition (0.77% carbon) and further cooling transforms the remaining austenite to pearlite. The resulting structure is a mixture of primary or pro-eutectoid ferrite (ferrite that formed above the eutectoid reaction) and regions of pearlite.

Hypereutectoid steels are steels that contain greater than the eutectoid amount of carbon. When such steel cools, as shown in z-z' of Figure 8-1 the process is similar to the hypo-eutectoid case, except that the primary or pro-eutectoid phase is now cementite instead of ferrite. As the carbon-rich phase forms, the remaining austenite decreases in carbon content, reaching the eutectoid composition at 727℃ (1341℉). As before, any remaining austenite transforms to pearlite upon slow cooling through this temperature.

It should be remembered that the transitions that have been described by the phase diagrams are for equilibrium conditions, which can be approximated by slow cooling. With slow heating, these transitions occur in the reverse manner. However, when alloys are cooled rapidly, entirely different results may be obtained, because sufficient time is not provided for the normal phase reactions to occur, in such cases, the phase diagram is no longer a useful tool for engineering analysis.

Hardening

Hardening is the process of heating a piece of steel to a temperature within or above its critical range and then cooling it rapidly. If the carbon content of the steel is known, the proper temperature to which the steel should be heated may be obtained by reference to the iron-iron carbide phase diagram. However, if the composition of the steel is unknown, a little preliminary experimentation may be necessary to determine the range. A good procedure to follow is to heat-quench a number of small specimens of the steel at various temperatures and to observe the result, either by hardness testing or by microscopic examination. When the correct temperature is obtained, there will be a marked change in hardness and other properties.

In any heat-treating operation the rate of heating is important. Heat flows from the exterior to the interior of steel at a definite rate. If the steel is heated too fast, the outside becomes hotter than the interior and uniform structure cannot be obtained. If a piece is irregular in shape, a slow rate is all the more essential to eliminate warping and cracking. The heavier the section, the longer must be the heating time to achieve uniform results. Even after the correct temperature has been reached, the piece should be held at that temperature for a sufficient period of time to permit its thickest section to attain a uniform temperature.

The hardness obtained from a given treatment depends on the quenching rate, the carbon content, and the work size. In alloy steels the kind and amount of alloying element influence only the hardenability (the ability of the workpiece to be hardened to depths) of the steel and does not affect the hardness except in unhardened or partially hardened steels.

Steel with low carbon content will not respond appreciably to hardening treatment. As the carbon content in steel increases up to around 0.60%, the possible hardness obtainable also increases. Above this point the hardness can be increased only slightly, because steels above the eutectoid point are made up entirely of pearlite and cementite in the annealed state. Pearlite responds best to heat-treating operations; any steel composed mostly of pearlite can be transformed into hard steel.

As the size of parts to be hardened increases, the surface hardness decreases somewhat even though all other conditions have remained the same. There is a limit to the rate of heat flow through steel. No matter how cool the quenching medium may be, if the heat inside a large piece cannot escape faster than a certain critical rate, there is a definite limit to the inside hardness. However, brine or water quenching is capable of rapidly bringing the surface of the quenched part to its own temperature and maintaining it at or close to this temperature. Under these circumstances there would always be some finite depth of surface hardening regardless of size. This is not true in oil quenching, when the surface temperature may be high during the critical stages of quenching.

Tempering

Steel that has been hardened by rapid quenching is brittle and not suitable for most uses. By tempering or drawing, the hardness and brittleness may be reduced to the desired point for service conditions. As these properties are reduced there is also a decrease in tensile strength and an increase in the ductility and toughness of the steel. The operation consists of reheating quench-hardened steel to some temperature below the critical range followed by any rate of cooling. Although this process softens steel, it differs considerably from annealing in that the process lends itself to close control of the physical properties and in most cases does not soften the steel to the extent that annealing would. The final structure obtained from tempering a fully hardened steel is called tempered martensite.

Tempering is possible because of the instability of the martensite, the principal constituent of hardened steel. Low-temperature draws, from 300℉ to 400℉ (150℃~205℃), do not cause much decrease in hardness and are used principally to relieve internal strains. As the tempering temperatures are increased, the breakdown of the martensite takes place at a faster rate, and at about 600℉(315℃) the change to a structure called tempered martensite is very rapid. The tempering operation may be described as one of precipitation and agglomeration or coalescence of cementite. A substantial precipitation of cementite begins at 600℉(315℃), which produces a decrease in hardness. Increasing the temperature causes coalescence of the carbides with continued decrease in hardness.

In the process of tempering, some consideration should be given to time as well as to temperature. Although most of the softening action occurs in the first few minutes after the temperature is reached, there is some additional reduction in hardness if the temperature is maintained for a prolonged time. Usual practice is to heat the steel to the desired temperature and hold it there only long enough to have it uniformly heated.

Two special processes using interrupted quenching are a form of tempering. In both, the hardened steel is quenched in a salt bath held at a selected lower temperature before being allowed to cool. These processes, known as austempering and martempering, result in products having certain desirable physical properties

Annealing

The primary purpose of annealing is to soften hard steel so that it may be machined or cold worked. This is usually accomplished by heating the steel too slightly above the critical temperature, holding it there until the temperature of the piece is uniform throughout, and then cooling at a slowly controlled rate so that the temperature of the surface and that of the center of the piece are approximately the same. This process is known as full annealing because it wipes out all trace of previous structure, refines the crystalline structure, and softens the metal. Annealing also relieves internal stresses previously set up in the metal.

The temperature to which a given steel should be heated in annealing depends on its composition; for carbon steels it can be obtained readily from the partial iron-iron carbide equilibrium diagram. When the annealing temperature has been reached, the steel should be held there until it is uniform throughout. This usually takes about 45min for each inch (25mm) of thickness of the largest section. For maximum softness and ductility the cooling rate should be very slow, such as allowing the parts to cool down with the furnace. The higher the carbon content, the slower this rate must be.

Normalizing and Spheroidizing

The process of normalizing consists of heating the steel about 50°F to 100°F (10°C~40°C) above the upper critical range and cooling in still air to room temperature. This process is principally used with low- and medium-carbon steels as well as alloy steels to make the grain structure more uniform, to relieve internal stresses, or to achieve desired results in physical properties. Most commercial steels are normalized after being rolled or cast.

Spheroidizing is the process of producing a structure in which the cementite is in a spheroidal distribution. If steel is heated slowly to a temperature just below the critical range and held there for a prolonged period of time, this structure will be obtained. The globular structure obtained gives improved machinability to the steel. This treatment is particularly useful for hypereutectoid steels that must be machined.

Surface Hardening

Carburizing

The oldest known method of producing a hard surface on steel is case hardening or carburizing. Iron at temperatures close to and above its critical temperature has an affinity for carbon. The carbon is absorbed into the metal to form a solid solution with iron and converts the outer surface into high-carbon steel. The carbon is gradually diffused to the interior of the part. The depth of the case depends on the time and temperature of the treatment. Pack carburizing consists of placing the parts to be treated in a closed container with some carbonaceous material such as charcoal or coke. It is a long process and used to produce fairly thick cases of from 0.03 to 0.16 in. (0.76~4.06mm) in depth.

Steel for carburizing is usually a low-carbon steel of about 0.15% carbon that would not in itself responds appreciably to heat treatment. In the course of the process the outer layer is converted into high-carbon steel with a content ranging from 0.9% to 1.2% carbon.

A steel with varying carbon content and, consequently, different critical temperatures requires a special heat treatment. Because there is some grain growth in the steel during the prolonged carburizing treatment, the work should be heated to the critical temperature of the core and then cooled, thus refining the core structure. The steel should then be reheated to a point above the transformation range of the case and quenched to produce a hard, fine structure. The lower heat-treating temperature of the case results from the fact that hypereutectoid steels are normally austenitized for hardening just above the lower critical point. A third tempering treatment may be used to reduce strains.

Carbonitriding

Carbonitriding, sometimes known as dry cyaniding or nicarbing, is a case-hardening process in which the steel is held at a temperature above the critical range in a gaseous atmosphere from which it absorbs carbon and nitrogen. Any carbon-rich gas with ammonia can be used. The wear-resistant case produced ranges from 0.003 to 0.030 inch, (0.08~0.76mm) in thickness. An advantage of carbonitriding is that the hardenability of the case is significantly increased when nitrogen is added, permitting the use of low-cost steels.

Cyaniding

Cyaniding, or liquid carbonitriding as it is sometimes called, is also a process that combines the absorption of carbon and nitrogen to obtain surface hardness in low-carbon steels that do not respond to ordinary heat treatment. The part to be case hardened is immersed in a bath of fused sodium cyanide salts at a temperature slightly above the Ac_1 range [6], the duration of soaking depending on the depth of the case. The part is then quenched in water or oil to obtain a hard surface. Case depths range from 0.005 to 0.015 inch. (0.13~0.38mm) may be readily obtained by this process. Cyaniding is used principally for the treatment of small parts.

Nitriding

Nitriding is somewhat similar to ordinary case hardening, but it uses a different material and treatment to create the hard surface constituents. In this process the metal is heated to a temperature of around 950°F (510°C) and held there for a period of time in contact with ammonia gas. Nitrogen from the gas is introduced into the steel, forming very hard nitrides that are finely dispersed through the surface metal.

Nitrogen has greater hardening ability with certain elements than with others; hence, special nitriding alloy steels have been developed. Aluminum in the range of 1% to 1.5% has proved to be especially suitable in steel, in that it combines with the gas to form a very stable and hard constituent. The temperature of heating ranges from 925°F to 1,050°F (495°C~565°C).

Liquid nitriding utilizes molten cyanide salts and, as in gas nitriding, the temperature is held below the transformation range. Liquid nitriding adds more nitrogen and less carbon than either cyaniding or carburizing in cyanide baths. Case thickness of 0.001 to 0.012in. (0.03~0.30mm) is obtained, whereas for gas nitriding the case may be as thick as 0.025 inch. (0.64mm). In general the uses of the two-nitriding processes are similar.

Nitriding develops extreme hardness in the surface of steel. This hardness ranges from 900 to 1,100 Brinell, which is considerably higher than that obtained by ordinary case hardening.

Nitriding steels, by virtue of their alloying content, are stronger than ordinary steels and respond readily to heat treatment. It is recommended that these steels be machined and heat-treated before nitriding, because there is no scale or further work necessary after this process. Fortunately, the interior structure and properties are not affected appreciably by the nitriding treatment and, because no quenching is necessary, there is little tendency to warp, develop cracks, or change condition in any way. The surface effectively resists corrosive action of water, saltwater spray, alkalies, crude oil, and natural gas.

◇ *New Words and Expressions*

corrosive [kəˈroʊsɪv]　　　　adj. 腐蚀性的；侵蚀性的
nitrogen [ˈnaɪtrədʒən]　　　　n. [化]氮，氮气
alloy [ˈælˌɔɪ]　　　　　　　　n. 合金
hardness [ˈhɑrdnɪs]　　　　　n. 坚硬，硬度
cyanide [ˈsaɪəˌnaɪd]　　　　　n. 氰化物
uniform [ˈjuːnɪfɔːrm]　　　　n. 制服，军服；adj. 一样的，规格一致的
crude oil　　　　　　　　　　原油
natural gas　　　　　　　　　天然气

⇨ *Notes*

[1] Alloy steel owe their properties to the presence of one or more elements other than carbon namely nickel, chromium, manganese, molybdenum, tungsten, silicon, vanadium, and copper.

参考译文：合金钢的性质取决于其所含有的除碳以外的一种或多种元素，如镍、铬、锰、钼、钨、硅、钒和铜。

Own M to N——M 归因于 N。

[2] Those portions of the iron-carbon diagram near the delta region and those above 20% carbon content are of little important to the engineer and are deleted.

参考译文：对工程人员来讲，铁-碳图中的近铁素体和含碳量大于 2% 的部分不重要，所以这两部分被去掉。

[3] The iron wants to change from the FCC austenite structure to the BCC ferrite structure, but the ferrite can only contain 0.02% carbon in solid solution.

参考译文：铁需要从面心立方奥氏体结构转变为体心立方铁素体结构，但是铁素体只能容纳固溶体状态的 0.02% 的碳。

FCC——face-centered cubic　面心立方
BCC——body-centered cubic　体心立方
in solid solution——固溶体状态

[4] This structure is composed of two distinct phases, but has its own set of characteristic

properties and goes by the name pearlite, because of its resemblance to mother-of-pearl at low magnification.

参考译文：这种结构由两种截然不同的状态组成，但它本身具有一系列特性，且因与低倍数放大时的珠母层有类同之处而被称为珠光体。

its own set of characteristic properties——本身有的一系列特性

goes by the name——称为

mother–of–pearl——母层 (珠光体层)

at low magnification——低倍放大时

[5] Tie-line and lever-law calculations show that low-carbon ferrite nucleates and grows, leaving the remaining austenite richer in carbon.

参考译文：由截线及杠杆定律分析可知，低碳铁素体结核并长大，剩下含碳量高的奥氏体。

tie-line——截线

lever-law——杠杆定律

[6] The part to be case hardened is immersed in a bath of fused sodium cyanide salts at a temperature slightly above the Ac_1 range…

参考译文：将需表面硬化的零件在略高于 Ac_1 range 温度熔化的氰化钠盐溶液中浸泡。

Ac_1——由珠光体转变为奥氏体的临界温度。

Questions

1. What is heat treatment?
2. Please explain metallurgy in your own words.
3. What is hardening?
4. Please explain carbonitriding in your own words.
5. Please explain liquid carbonitriding in your own words.

Reading Material: Heat Treatment of Steels

Heat treating refers to the heating and cooling operations performed on a metal for the purpose of altering such characteristics as hardness, strength, or ductility. A tool steel intended to be machined into a punch may first be softened so that it can be machined[1]. After being machined to shape, it must be hardened so that it can sustain the punishment that punches receive. Most heating operations for hardening leave a scale on the surface, or contribute other surface defects. The final operation must, therefore, be grinding to remove surface defects and provide a suitable surface finish.

When a steel part is to be either hardened or softened, its temperature must be taken above the critical temperature line; that is, the steel must be austenitized. Usually a temperature of 50 to

100 degree above the critical temperature is selected, to ensure that the steel part reaches a high enough temperature to be completely austenitized, and also because furnace temperature control is always a little uncertain[2].

The steel must be held at furnace temperature for sufficient time to dissolve the carbides in the austenite, after which the steel can be cooled. How much residence time in the furnace is required is to some degree a matter of experience with any particular steel.

Usually, for a 3/4 in. bar (1in=0.0254m), 20 minutes or slightly more will do. Double the time for twice the diameter. Alloy steels may require a longer furnace time; many of these steels are best preheated in a lower—temperature furnace before being charged into the hardening furnace.

When the heating time is completed, the steel must be cooled down to room temperature. The cooling method determines whether the steel will be hardened or softened. If the steel is quickly removed from the furnace and quenched into cold water, it will be hardened. If it is left in the furnace to cool slowly with the heat turned off, or cooled in air (small pieces of plain carbon steel cannot be air—softened, however), it will be softened. High-alloy steels may be hardened by air-cooling, but plain carbon steels must have a more severe quench, almost always water.

There are several softening methods for steels, and the word softening therefore does not indicate what softening process or purpose was used. The method of softening by slow cooling from austenite is called annealing, not softening. Annealing leaves the steel in the softest possible condition (dead soft).

To conclude, the difference between hardening and annealing is not in the heating process but in the cooling process.

◈ *New Words and Expressions*

treatment ['triːtmənt]	n. 处理，加工，对待
refer(to) [rɪˈfɚ]	v. 指的是，涉及，查阅，参考
cool [kuːl]	adj. & v. 冷的，凉的，冷却
ductility [dʌkˈtɪlɪtɪ]	n. 可延性，韧性
machine [məˈʃiːn]	v. 机械加工，切削加工
punch [pʌntʃ]	n. & v. 冲头，冲床，冲孔器；冲压
soften [ˈsɒfn]	v. 软化，变软
harden [ˈhɑːdn]	v. 硬化，变硬，淬硬
sustain [səˈsteɪn]	v. 承受得住，支撑，维持，遭受
punishment [ˈpʌnɪʃmənt]	n. 大负荷，损伤，惩罚
scale [skeɪl]	n. & v. 氧化皮，刻度，起氧化皮
contribute [kənˈtrɪbjuːt]	v. 产生，促使，提供，贡献
defect [ˈdiːfekt]	n. 缺陷，疵点

grind [graɪnd]	v.	磨，磨削，研磨
suitable ['su:təbl]	adj.	适宜的，相适应的
finish ['fɪnɪʃ]	v. & n.	结束，精加工，粗糙度
austenitize [ɔ:s'tenətaɪz]	n.	使成奥氏体，奥氏体化
select [sɪ'lekt]	v.	选择，挑选
ensure [ɛn'ʃʊr]	v.	确保，保证得到
uncertain [ʌn'sɜ:tn]	n.	不定的，不可靠的，易变的
carbide ['kɑ:baɪd]	n.	碳化物，硬质合金
austenite	n.	奥氏体
residence ['rezɪdəns]	n.	停留
bar [bɑ:(r)]	n.	棒，棒料，杆，条
slight(~ly) [slaɪt]	adj. & adv.	轻微的，稍微，轻微地
double(~ly) ['dʌbl]	adj.	两倍的，双重地，加倍，两倍
twice [twais]	adv.	两倍，两次
diameter [dai'mmitə]	n.	直径
preheat [pri'hi:t]	v.	预热
method ['mɛθəd]	n.	方法，方式
quench [kwentʃ]	v. &n.	淬火
plain [plein]	adj.	普通的，平常的，简易的
severe(~ly) [silviə]	adj. &adv.	剧烈的(地)，严格的(地)
indicate ['indikeit]	v.	表明，指示
anneal [ə'ni:1]	v.	退火
soft [sɔft]	adj.	软的
conclude [kən'klud]	v.	结束，断定
difference ['dɪfərəns,]	n.	差异，区别，不同
heat treatment		热处理
for the purpose of		为了，以便
critical temperature		临界温度
to some degree		在某种程度上
room temperature		常温，室温
dead soft		极软
to conclude		最后(一句话)

⇨ **Notes**

[1] A tool steel intended to be machined into a punch may first be softened so that it can be machined.

参考译文：要想把一块工具钢加工成一只冲头，可先使其变软，以便能进行机械加工。

本句为主从复合句。so that 引导的是目的状语从句，主句中 intended to be machined into a punch 是分词短语，相当于 which is intended to be machined into a punch，用来修饰 a tool steel。intend 可用于 SVOC 句型，用 to do 作补足语，变成被动语态时就成为 SV(be+ intended to) C 句型。例如：

We intended a tool steel to be machined into a punch.
译文：我们打算把一块工具钢加工成一只冲头。
A tool steel was intended to be machined into a punch.
译文：打算把一块工具钢加工成一只冲头。

[2] Usually a temperature of 50 to 100 degree above the critical temperature is selected, to ensure that the steel part reaches a high enough temperature to be completely austenitized, and also because furnace temperature control is always a little uncertain.

参考译文：通常，所选择的温度要比临界温度高出 50℃～100℃，以确保钢件达到足够高的温度，从而完全奥氏体化，同时也由于炉温控制总会有点偏差。

本句为主从复合句，that 引导的是宾语从句，because 引导的是原因状语从句。句中 high enough… to be completely austenitized 是一个割裂的形容词短语，用来修饰 temperature。本句可写成…a temperature high enough to be completely austenitized…。

Questions

1. What is the disadvantage of heat treatment?
2. What is the process of heat treatment?
3. When must the steel be cooled down to room temperature?

Unit 9 Welding

The American Welding Society defines welding[1] as "a localized coalescence[2] of metals wherein coalescence is produced by heating to suitable temperatures with or without the application of pressure and with or without the use of filler metal. The filler[3] metal either has a melting point approximately the same as the base metals or has a melting point below that of the base metals but above 800℉."

There are 34 different welding processes. Figure 9-1 is a master chart of these processes and shows their connection to major welding methods. The proper choice of a particular method of weld must be carefully considered by the designer. And, before making a final selection, the designer will have to consider, evaluate, and weigh such factors as the metals to be joined, the joint design, the thicknesses (or bulk) of metals, the type of load, the equipment available, the production rate, and the environment to which the weld will be subjected. It is apparent that there are no "hard and fast" rules that one can use in making a decision with the exception, perhaps, of a particular case where one requirement or condition is of overall importance.

Because of space restrictions, it is not possible to discuss in detail each of the processes in Figure 9-1. We shall confine ourselves to describing briefly the most widely used welding methods.

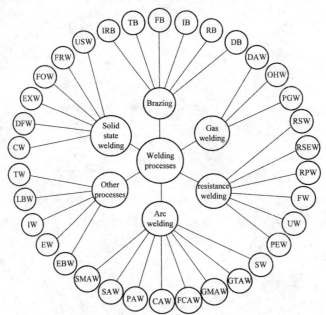

Figure 9-1 Master chart of welding processes

Gas Welding

Gas welding is "a group of welding processes wherein coalescence is produced by heating with a gas flame or flames with or without the application of pressure and with or without the use of filler material."

Of the three types of gas welding, oxyacetylene welding (OAW) is the one most frequently employed. This method uses a mixture of oxygen and acetylene to production heating. Fluxes[4] may be used to reduce oxidation and to promote a better welding joint. This type of welding is suitable for both ferrous (including cast iron) and nonferrous metals and is capable of welding thick metal sections.

Oxyhydrogen welding (OHW) is used for low melting point metals such as aluminum magnesium and lead.

Pressure gas welding (PGW) uses an oxyacetylene flame for a heat source but does not require a filler rod. Instead, fusion is obtained by applying pressure to the heated parts, either while being heated or after the parts are heated. This form of welding can be used for joining both ferrous and nonferrous metals.

Arc Welding

Arc welding is "a welding process wherein coalescence is produced by heating with an arc or arcs with or without the application of pressure and with or without the use of filler metals." As indicated in Figure 9-1, there are eight different arc welding processes. These are (1) carbon-arc welding, (2) shielded metal-arc welding, (3) flux cored arc welding, (4) gas metal-arc welding, (5) gas tungsten-arc welding, (6) submerged arc welding, (7) plasma-arc[5] welding, and (8) stud[6] welding.

The most widely used of these methods is the shielded metal-arc welding (SMAW) process. It is defined as an arc welding process wherein coalescence is produced by heating with an arc between a covered metal electrode[7] and the work. Shielding is obtained from decomposition of the electrode covering[8]. Pressure is not used and filler metal is obtained from the electrode.

The shielded metal-arc process is employed in both manual and automated production setups. Electrodes are available that permit the welding of ferrous metals (including cast iron), all grades of carbon steels, low-alloy high-strength steels, stainless steels, copper bearing steels, copper alloys, aluminum, nickel, nickel alloys, and bronze. This welding technique is used in many fields, particularly in the manufacture of machinery, transportation equipment, piping systems and in various structures (for example, building, trusses, machine bases, and so on).

The next two most widely used arc welding methods are the submerged arc welding (SAM) process and the plasma-arc welding (PAW) process.

Submerged arc welding (SAW) is "an arc welding process wherein coalescence is produced by heating with an arc or arcs between a bare metal electrode or electrodes and the work. The arc is shielded by a blanket of granular, fusible material on the work. Pressure is not used and filler metal is obtained from the electrode and sometimes from a supplementary welding rod."

This method can be used in fully automated equipment where the feeds of both the electrode and granular flux are controlled. The method is also adaptable for semiautomatic equipment where the feeds of the electrode and granular flux are controlled manually. Since the granular flux must cover the joint to be welded, this method is restricted to parts in horizontal position and is particularly suited for welding long straight joints. Also, fewer passes are needed to weld thick metal sections than are usually required by shielded metal-arc welding.

Submerged arc welding can be used to weld low carbon steels, high-strength low-alloy steels, chromium steels and austenitic chromium-nickel steels. With special methods, it is also possible to weld high-alloy air-hardening steels.

Plasma-arc welding(PAW) is "an arc welding process wherein coalescence is produced by heating with a constricted arc between an electrode and the workpiece (transferred arc) or the electrode and the constricting nozzle[9] (non-transferred arc). Shielding is obtained from the hot, ionized gas issuing from the orifice, which may be supplemented by an auxiliary source of shielding gas. Shielding gas may be an inert gas or a mixture of gases. Pressure may or may not be used, and filler may or may not be supplied.

Plasma-arc welding is used for quality welding and can easily weld 5-inch, thick aluminum sections or stainless steel sections up to 4 inch thick. Since there are no products of combustion, the welded joints have no porosity and display a strong resistance to high stresses and impact loading.

The plasma torch is constructed with an electrode centrally within a metal cup that guides an inert streaming gas past the electrode. In the plasma-arc torch, the discharge end of the cup is smaller in diameter than the upper diameter so that a discharge nozzle is created. In addition, the inner wall of the nozzle is lined with a ceramic material.

Although a plasma stream can be created with any gas, a gas that is non-oxidizing should be used. Another important requirement is the thermal conductivity of the gas rather than the temperature it may attain. Thus, gases of higher conductivity can transfer more heat, making it possible to weld "bulky" sections more easily. Argon, helium, and hydrogen are the gases most frequently used. Hydrogen has the higher thermal conductivity and produces hotter arcs than those produced by argon or helium.

Other forms of arc welding, namely, carbon-arc welding (CAW), flux cored arc welding (FCAW), gas metal-arc welding (GMAW), gas tungsten-arc welding (GTAW) and stud welding (SW) are used for joining particular metals or for mass production. For example, carbon-arc welding is used to join galvanized sheet steel, brass, bronze, and aluminum, whereas flux cored arc welding uses a flux cored electrode, continuously fed from a spool for quantity production.

Stud welding is accomplished by means of a stud-welding gun, which welds a stud to the surface of a workpiece. The method is extensively employed in the automotive, shipbuilding, railroad, and building construction industries.

◇ New Words and Expressions

localized coalescence	局部熔结 (聚结、结合)
filler metal	填料金属
oxyacetylene elding(OAW)	氧乙炔焊
oxyhydrogen welding (OHW)	氢氧焊
pressure gas welding (PGW)	加压气焊
ferrous metal	黑色金属 (含铁金属)
non-ferrous metal	有色金属
carbon-arc welding	碳弧焊
shielded metal-arc welding	保护金属电弧焊
flux cored arc welding	焊芯电弧焊
gas metal-arc welding	气保护金属级电弧焊
gas tungsten-arc welding	气保护钨极电弧焊
submerged welding	埋弧焊
plasma-arc welding	等离子焊接
stud welding	电栓焊，螺柱焊接
covered metal electrode	覆盖耀皮的金属焊条
transferred arc	传导电弧
non-transferred arc	非传导电弧
constricted arc	压缩电弧
ionized gas	电离气体
discharge end of the cup	金属环的出气 (出料，放料) 端

⇨ Notes

[1] welding: a localized coalescence of metals wherein coalescence in produced by heating to suitable temperature with or without the application of pressure and with or without the use of filler metal.

参考译文：焊接：一种局部的金属的接合，将金属加热到适当的温度而使之产生接合，可加压力也可不加压力，可用填充金属也可不用填充金属。

[2] coalescence: to unit as a whole.

参考译文：接合：成为一个整体。

[3] filler: a substance added in a product as to increase bulk, weight, viscosity, opacit, or strength.

参考译文：填料：加到产品中的一种物质，用以增强其体积，重量，黏性，不透明度或强度。

[4] flux: a substance (as rosin) applied to surfaces to be joined by soldering, brazing, or welding to clean and free them from oxide and promote union.

参考译文：焊剂：如松香等一类物质，用到锡焊，钎焊或焊接的结合面上，以去除氧化物，改善结合强度。

[5] plasma: a collection of charged particles (as in the atmospheres of stars or in a metal) containing about equal numbers of positive ions and electrons and exhibiting some properties of a gas but differing from a gas in being a good conductor of electricity and in being affected by a magnetic field.

参考译文：等离子体：带电离子流（如在恒星的大气层中或金属中），其中正离子和电子的数目相等，性质像气体，但又与气体不同，等离子体是电的良导体，受磁场的影响会产生偏转。

[6] stud: any of various in-fixed pieces (as a rod or a pin) projecting from a machine and serving chiefly as a support or axis.

参考译文：柱栓：各种嵌入件（如杆或销），一端从机器上突出来，主要用于支撑或作轴心。

[7] electrode: either pole (anode, cathode) of an electric battery, or a conductor used to establish electric contact with a nonmetallic part of a circuit.

参考译文：电极：电池的任意一个极（阳极或阴极），或者是一个导体用以和电路中的一个非金属件建立通电的路径。

[8] Shielding is obtained from decomposition of the electrode covering.

参考译文：焊条药皮的分解形成了保护。

[9] nozzle: a short tube with a taper of constriction used (as on a hose) to speed up or direct a flow of fluid.

参考译文：喷嘴：锥状或颈缩状的短管（如水管），使液体的流动加速或对之进行导向。

Questions

1. What is the definition of welding according to American Welding Society?
2. What is gas welding?
3. What is plasma-arc welding (PAW)?
4. What is submerged arc welding (SAW)?
5. What is pressure gas welding (PGW)?

Reading Material: Welding

Resistance Welding

Resistance welding is "a group of welding processes wherein coalescence is produced by the heat obtained from resistance of the work to electric current in a circuit of which the work in a part, and by the application of pressure." Figure 9-1 indicates that there are six types of resistance welding processes. They are (1) resistance spot welding, (2) resistance seam welding, (3) projection welding, (4) flash welding, (5) upset welding, and (6) percussion welding. Resistance welding is widely used in quantity production. By means of proper controls and tooling, it is readily adaptable to automation, including any required preheating or heat treatment after welding. The most widely used types of resistance welding are the spot, seam, and projection forms of welding.

Resistance spot welding (RSW) "A resistance welding process wherein coalescence at the faying surfaces is produced in one spot by the heat obtained from the resistance to electric current through the work parts held together under pressure by electrodes. The size and shape of the individually formed welds are limited primarily by the size and contour of the electrodes."

Spot welding is primarily restricted to thin metals (for example, 0.001 in. thick to 0.125 in. thick for steel and magnesium, 0.156 in. thick for aluminum), namely, steels, stainless steels, aluminum, magnesium, nickel alloys, bronze, and brass. Some dissimilar metals can be spot welded, but with difficulty.

Resistance seam welding (RSEW) "A resistance welding process wherein coalescence at the faying surfaces is produced by the heat obtained from the resistance to electric current through the work parts held together under pressure by electrodes. The resulting weld is a series of overlapping resistance-spot welds made progressively along a joint by rotating the electrodes." In principle, seam welding is similar to spot welding except that the weld is continuous by virtue of the rollers rather than discontinuous as in spot welding.

Seam welding is primarily used for quantity production but is restricted to joining metal gages that are thinner than those which can be joined by spot welding. The "normal" range of thicknesses compatible with seam welding is 0.100-0.125 inch.

Projection welding (PW) "A resistance welding process wherein coalescence is produced by the heat obtained from the resistance to electric current through the work parts held together under pressure by electrodes. The resulting welds are localized at predetermined points by projections, embossments, and intersections."

Projection welding is a process similar to spot welding except that the projections tend to localize the heat, permitting thicker materials to be welded. Simultaneous welds are readily made by this method, and result in a stronger welded structure than that obtained with spot welding.

Flash welding (FW) In this process abutting surfaces to be welded are clamped in fixtures and brought within close proximity (or light contact) of each other so that an electric is produced between the surfaces causing them to heat to a fusible temperature. At this point, the two surfaces are forced together, completing the weld. Forcing the two surfaces together causes the metal to be displaced (that is, bulge) outward from the welded joint. This "upset" metal is usually removed after welding. Preheating (for large bulky parts) and postheating (that is heat treatment) can readily be made part of the overall welding cycle.

Materials that are easily weldable by spot welding are also weldable by flash welding, although the method is used mostly with ferrous metals. Copper, copper alloys, and some aluminum cannot be relied upon to produce satisfactorily welded joints. However dissimilar metals can readily be welded by this method, including even refractory metals such as tungsten, molybdenum, and tantalum.

Upset welding (UW) A process similar to flash welding except the parts to be welded are held in close contact with each other before the electric circuit is closed. Thus, there is no flashing in this method. Upset welding is extensively used in the fabrication of tubular sections, pipe, and heavy steel rings; it is also used for joining small ferrous and nonferrous strips.

Percussion welding (PEW) "A resistance welding process wherein coalescence is produced simultaneously over the entire abutting surfaces by heat obtained from an arc produced by a rapid discharge of electrical energy with pressure percussively applied during or immediately following the electrical discharge."

Percussion welding is used to for special joining situations (for example, joining dissimilar metals that cannot be welded economically by flash welding). This welding method is also used to weld pins, studs, and bolts, and so on, to other components as well as to join sections of pipe, rod, or tube to each other or to flat sections.

Brazing

Brazing is "a welding process wherein coalescence is produced by heating to suitable temperatures and by using a filler metal having a liquidus above 427°C and below the solidus of the base metals. The filler metal is distributed between the closely fitted surfaces of the joint by capillary attraction." As indicated by Figure 9-1, there are six brazing methods, namely (1) infrared brazing (IRB), (2) torch brazing(TB), (3) furnace brazing(FB), (4) induction brazing (IB), (5) resistance brazing(RB), (6) dip brazing (DB). Among these methods, the primary difference is the manner in which the metal to be joint are heated. Also, only four of the six methods are of industrial importance, torch brazing, furnace brazing, induction brazing, and dip brazing. These methods are defined and briefly described in the following paragraphs.

Torch brazing A joint process that may employ acetylence, natural gas, butane, or propane in combination with air or oxygen to supply the heat required to melt the filler rod and diffuse it into the surface of the base metal. This technique is not extensively used for continuous mass production.

Furnace brazing A high production fabrication method where the heat is supplied by gas or electric heating coils. The furnace is of the box type or the continuous type, which employ a wire mesh belt to transport the part to be brazed. Furnace brazing requires that preformed shapes of filler metal be placed on the parts to be jointed prior to entering the furnace. This method of brazing is well suited to high production and can avoid the use of fluxing by maintaining an inert atmosphere in the furnace.

Induction brazing Like furnace brazing, induction brazing requires the use of preformed shapes of filler metal. Heat is produced by placing the parts to be brazed within the field of a high frequency induction coil. The workpieces are heated by eddy currents because the parts to be jointed offer electric-magnetic resistance to the changing induction field. Heating is very rapid, and by properly shaping induction coil, the heat can be applied in the local area of the joint to be brazed. Induction brazing can be used for mechanized production when properly designed tooling and feeding devices are incorporated into the production setup.

Dip brazing Parts can be dip brazed by one of two methods. In chemical dip brazing, the parts jointed are prepared with preformed filler metal, after which they are placed into a molten bath brazing flux. In the molten metal bath process, the assembled parts are first refluxed and then immersed into a molten bath of filler metal. This latter method of brazing is restricted to small parts, whereas the former method is more adaptable for joining large parts.

Other Welding Processes

The master chart of welding processes (Figure 9-1) shows a series of solid state welding techniques and some other processes. These methods are defined by the American Welding Society as following:

Ultrasonic welding (USW) A solid state welding process wherein coalescence is produced by the local application of high frequency vibratory energy as the work parts are held together under pressure.

Friction welding (FRW) A solid state welding process wherein coalescence is produced by heat obtained from mechanically induced sliding motion between rubbing surfaces. The parts are held together under pressure.

Forge welding (FOW) A solid state welding process wherein coalescence is produced by heating and by applying pressure to cause permanent deformation at the interface.

Explosion welding (EXW) A solid state welding process wherein coalescence is effected by high velocity movements produced by a controlled detonation.

Diffusion welding (DFW) A solid state welding process wherein coalescence of the faying surfaces is produced by the application of pressure and elevated temperatures. The process does not involve macroscopic deformation or relative motion of parts. A solid filler metal may be or may not be inserted.

Cold welding (CW) A solid state welding process wherein coalescence is produced by the external application of mechanical force alone.

Thermit welding (TW) A group of welding processes wherein coalescence is produced by heating with superheated liquid metal and slag resulting from a chemical reaction between a metal oxide and aluminum with or without the application of pressure. Filler metal, when used, is obtained from the liquid metal.

Induction welding (IW) A welding process wherein coalescence is produced by the heat obtained from the resistance of the work to induced electric current with or without the application of pressure.

Electroslag welding (ESW) A welding process wherein coalescence is produced by molten slag which melts the filler metal and the surfaces of the work to be welded. The weld pool is shielded by this slag which moves along the full cross section of the joint as welding progresses. The conductive slag is maintained molten by its resistance to electric current passing between the electrode and the work.

Electron beam welding (EBW) A welding process wherein coalescence is produced by the heat obtained from a concentrated beam composed primarily of high velocity electrons impinging upon the surfaces to be joined.

◇ New Words and Expressions

thermit welding(TW)	热剂焊
laser beam welding(LBW)	激光束焊接
electroslag welding(ESW)	电渣焊
electron beam welding(EBW)	电子束焊
resistance spot welding(RSW)	电阻点焊
faying surface	接合面
resistance seam welding(RSEW)	电阻缝焊 (滚焊)
by virtue of	借助于
projection welding(PW)	凸焊
flash welding(FW)	闪光焊
abutting surfaces	对接表面
upset welding(UW)	电阻对焊
percussion welding(PEW)	储能焊，冲击焊
brazing	钎焊 (硬钎焊，铜焊，硬焊)
capillary attraction	毛细作用
infrared brazing(IRB)	红外线钎焊
torch brazing(TB)	焊炬钎焊
furnace brazing(FB)	炉中钎焊
induction brazing(IB)	感应钎焊
resistance brazing(RB)	电阻加热 (接触) 钎焊
dip brazing(DB)	浸渍钎焊
eddy current	涡流
molten bath	熔池
preluxed	预加焊剂的
ultrasonic welding(USW)	超声波焊接
friction welding(FRW)	摩擦焊
forge welding(FOW)	锻焊
explosion welding(EXW)	爆炸焊

| diffusion welding(DFW) | 扩散焊 |
| cold welding(CW) | 冷压焊 |

Questions

1. What is the definition of resistance welding?
2. What is resistance spot welding (RSW)?
3. What are the welding processes?
4. What is electron beam welding (EBW)?
5. What is the contribution of American Welding Society?

Unit 10 Die Casting

In die casting, molten metal is forced by pressure into a metal mold known as a die. Because the metal solidifies under a pressure from 80 to 40,000 psi (0.6~275MPa), the casting conforms to the die cavity in shape and surface finish. The usual pressure is from 1500 to 2000 psi (10.3~14MPa).

Die casting is the most widely used of the permanent-mold processes. Two methods are employed: (1) hot chamber and (2) cold chamber.

A diagrammatic sketch illustrating the operation of horizontal-plunger cold-chamber machines is shown in Figure 10-1.

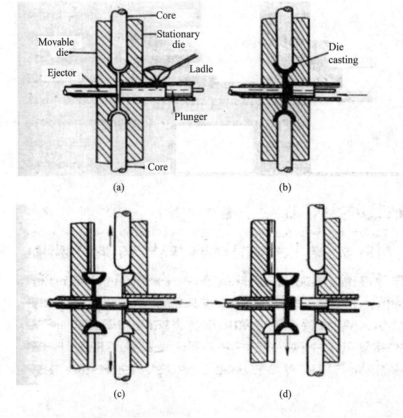

Figure 10-1 The die casting of brass, aluminum, or magnesium in horizontal-plunger cold-chamber machine

In Figure10-1(a) the dies are shown closed with cores in position and the molten metal ready to be poured. As soon as the ladle is emptied the plunger moves to the left and forces the metal into the mold (Figure10-1(b)). After the metal solidifies, the cores are withdrawn and the dies are opened. In Figure10-1(c) the dies are open and the casting is ejected from the stationary half. To complete the process of opening, the ejector rod comes into operation and ejects the casting from the movable half of the die ((Figure10-1(d)). This operating cycle is used in a variety of machines that operate at pressures ranging from 5600 to 22,000 psi (39 ~150MPa). These machines are fully hydraulic and semiautomatic. After the metal is poured, the rest of the operations are automatic. The injection of liquid metal under pressure into a mould offers several advantages. It means that very thin sections can be cast, that metal can be forced into the recesses of a mould of very complex shape, and that rapid solidification under pressure will considerably reduce porosity. Also, the need to provide risers in the mould is eliminated. In die casting there is little metal wastage and a casting is obtained that requires little or no machining. The process is, however, restricted to the lower melting point metal and alloys.

◇ *New Words and Expressions*

die casting	压力铸造
solidify [sə'lɪdɪfaɪ]	v. 凝固
pis=pounds per square inch	磅/平方英寸
finish ['fɪnɪʃ]	n. 光洁度
chamber ['tʃeɪmbə]	n. 室，腔
diagrammatic [ˌdaɪəɡrə'mætɪk]	adj. 图示的，概略的
plunger ['plʌn(d)ʒə]	n. 柱塞，活塞，插棒
ladle ['leɪd(ə)l]	n. 铸勺
eject [ɪ'dʒekt]	v. 推出，喷射
stationary ['steɪʃ(ə)n(ə)rɪ]	adj. 静止的，固定的
ejector [i'dʒektə]	n. 推顶器，推顶装置
movable ['muːvəb(ə)l]	adj. 可移动的
rang from	从……到……范围
semiautomatic ['semɪˌɔːtə'mætɪk]	adj. 半自动的
recess [rɪ'ses; 'riːses]	n. 凹入处，幽深处
porosity [pɔː'rɒsɪtɪ]	n. 孔隙，多孔性
eliminate [ɪ'lɪmɪneɪt]	vt. 除去，省去
wastage ['weɪstɪdʒ]	n. 浪费

✕ *Questions*

1. Which categories do the mechanical properies mainly fall into?
2. What is the definition of tensile strength?

3. What is hardness?
4. What is ductility?
5. What is toughness?

Reading Material: Sand Casting

Most metal castings are made by pouring molten metal into a prepared cavity and allowing it to solidify. The process dates from antiquity. The largest bronze statue in existence today is the great Sun Buddha in Nara, Japan. Cast in the eighth century, it weighs 551 tons (500 metric tons) and is more than 71 ft (21m) high. Artisans of the Shang Dynasty in China (1766-1222 B.C.) created art works of bronze with delicate filigree as sophisticated as anything that was designed and produced today.

There are many casting processes available today, and selecting the best one to produce a particular part depends on several basic factors, such as cost, size, production rate, finish, tolerance, section thickness, physical-mechanical properties, intricacy of design, machinability, and weldability.

Sand casting, the oldest and still the most widely used casting process, will be presented in more detail than the other processes since many of the concepts carry over into those processes as well.

Green Sand

Green sand generally consists of silica sand and additives coated by rubbing the sand grains together with clay uniformly wetted with water[1]. More stable and refractory sands have been developed, such as fused silica, zircon, and mullite, which replace lower-cost silica sand and have only 2% linear expansion at ferrous metal temperatures. Also, relatively unstable water and clay bonds are being replaced with synthetic resins, which are much more stable at elevated temperatures.

Green sand molding is used to produce a wide variety of castings in sizes of less than a pound to as large as several tons. This versatile process is applicable to both ferrous and nonferrous materials.

Green sand can be used to produce intricate molds since it provides for rapid collapsibility; that is, the mold is much less resistant to the contraction of the casting as it solidifies than are other molding processes. This results in less stress and strain in the casting.

The sand is rammed or compacted around the pattern by a variety of methods, including hand or pneumatic-tool ramming, jolting (abrupt mechanical shaking), squeezing (compressing the top and bottom mold surfaces), and driving the sand into the mold at high velocities (sand slinging). Sand slingers are usually reserved for use in making very large castings where great volumes of sand are handled.

For smaller castings, a two-part metal box or flask referred to as a cope and drag is used. First the pattern is positioned on a mold board, and the drag or lower half of the flask is positioned over it. Parting powder is sprinkled on the pattern and the box is filled with sand. A jolt squeeze machine quickly compacts the sand. The flask is then turned over and again parting powder is dusted on it. The cope is then positioned on the top half of the flask and is filled with sand, and the two-part mold with the pattern board sandwiched in between is squeezed.

Patterns

Patterns for sand casting have traditionally been made of wood or metal. However, it has been found that wood patterns change as much as 3% due to heat and moisture. This factor alone would put many castings out of acceptable tolerance for more exacting specifications. Now, patterns are often made from epoxies and from cold-setting rubber with stabilizing insets. Patterns of simple design, with one or more flat surface, can be molded in one piece, provided that they can be withdrawn without disturbing the compacted sand. Other patterns may be split into two or more parts to facilitate their removal from the sand when using two-partflasks. The pattern must be tapered to permit easy removal from the sand. The taper is referred to as draft. When a part does not have some natural draft, it must be added. A more recent innovation in patterns for sand casting has been to make them out of foamed polystyrene that is vaporized by the molten metal. This type of casting, known as the full-mold process, does not require pattern draft.

Sprues, Runners and Gates

Access to the mold cavity for entry of the molten metal is provided by sprues, runners, and gates, as shown in Figure 10-2. A pouring basin can be carved in the sand at the top of the sprue, or a pour box, which provides a large opening, may be laid over the sprue to facilitate pouring. After the metal is poured, it cools most rapidly in the sand mold. Thus the outer surface forms a shell that permits the still molten metal near the center to flow toward it. As a result, the last portion of the casting to freeze will be deficient in metal and, in the absence of a supplemental metal-feed source, will result in some form of shrinkage[2]. This shrinkage may take the form of gross shrinkage (large cavities) or the more subtle microshrinkage (finely dispersed porosity). These porous spots can be avoided by the use of risers, as shown in Figure 10-2, which provides molten metal to make up for shrinkage losses.

Cores

Cores are placed in molds wherever it is necessary to preserve the space it occupies in the mold as a void in the resulting castings[3]. As shown in Figure 10-2, the core will be put in place after the pattern is removed. To ensure its proper location, the pattern has extensions known as core prints that leave cavities in the mold into which the core is seated. Sometimes the core may be molded integrally with the green sand and is then referred to as a green-sand core. Generally, the core is made of sand bonded with core oil, some organic bonding materials, and water. These materials are thoroughly blended and placed in a mold or core box. After forming, they are removed and baked at 350°F to 450°F (177°C to 232°C). Cores that consist of two or more parts are pasted together after baking.

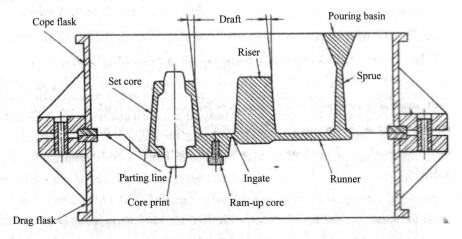

Figure 10-2 Sectional view of a casting mold

CO_2 Cores

CO_2 cores are made by ramming up moist sand in a core box. Sodium silicate is used as a binder, which is quickly hardened by blowing CO_2 gas over it. The CO_2 system has the advantage of making the cores immediately available.

Pouring the Metal

Several types of containers are used to move the molten metal from the furnace to the pouring area. Large castings of the floor-and-pit type are poured with a ladle that has a plug in the bottom, or, as it is called, a bottom-pouring ladle. It is also employed in mechanized operations where the molds are moved along a line and each is poured as it is momentarily stopped beneath the large bottom-pour ladle.

Ladles used for pouring ferrous metals are lined with a high alumina-content refractory. After long use and oxidation, it can be broken out and replaced. Ladles used in handling ferrous metals must be preheated with gas flames to approximately 2600°F to 2700°F (1427°C to 1482°C) before filling. Once the ladle is filled, it is used constantly until it has been emptied.

For nonferrous metals, simple clay-graphite crucibles are used. While they are quite susceptible to breakage, they are very resistant to the metal and will hold up a long time under normal conditions. They usually do not require preheating, although care must be taken to avoid moisture pickup. For this reason they are sometimes baked out to assure dryness.

The pouring process must be carefully controlled, since the temperature of the melt greatly affects the degree of liquid contraction before solidification, the rate of solidification, which in turn affects the amount of columnar growth present at the mold wall, the extent and nature of the dendritic growth, the degree of alloy burnout, and the feeding characteristics of the risering system[4].

Finishing Operations

After the castings have solidified and cooled somewhat, they are placed on a shakeout table or grating on which the sand mold is broken up, leaving the casting free to be picked out. The casting is then taken to the finishing room where the gates and risers are removed. Small gates and risers may be broken off with a hammer if the material is brittle. Larger ones require sawing, cutting with a torch, or shearing. Unwanted metal protrusions such as fins, bosses, and small portions of gates and risers need to be smoothed off to blend with the surface. Most of this work is done with a heavy-duty grinder and the process is known as snagging or snag grinding. On large castings it is easier to move the grinder than the work, so swing-type grinders are used. Smaller castings are brought to stand-or bench-type grinders. Hand and pneumatic chisels are also used to trim castings. A more recent method of removing excess metal from ferrous castings is with a carbon-air torch. This consists of a carbon rod and high-amperage current with a stream of compressed air blowing at the base of it. This oxidizes and removes the metal as soon as it is molten. In many foundries this method has replaced nearly all chipping and grinding operations.

◇ New Words and Expressions

casting ['kastɪŋ]	n. 铸造，铸件
cavity ['kæviti]	n. 空腔，型腔
solidify [sə'lidifai]	vt. 凝固
antiquity [æn'tikwiti]	n. 古代
buddha ['budə]	n. 佛
nara ['naːraː]	n. 奈良市
artisan [aːti'zæn]	n. 工匠
filigree ['filigriː]	n. 精细之作
finish ['finiʃ]	n. 光洁度
tolerance ['tɔlərəns]	n. 公差
intricacy ['intrɪkəsi]	n. 复杂
machinability [məʃiːnə'bilətɪ]	n. (可) 切削性
weldability [weldə'bilili]	n. (可) 焊接性
silica ['sɪlɪkə]	n. 石英
additive ['æditiv]	n. 添加剂
clay [klei]	n. 黏土
refractory [rɪ'fræktəri]	adj. 难熔的，耐火的

fuse [fju:z]	vt. 使熔化
zircon ['zə:kən]	n. 锆石
mullite ['mʌlait]	n. 富铝红柱石
ferrous ['ferəs]	adj. 铁的
resin ['rezin]	n. 树脂，松香
molding ['məuldɪŋ]	n. 铸型，造型
versatile ['və:sətail]	adj. 多用途的
nonferrous [nɔn'ferəs]	adj. 非铁的
intricate ['ɪntrikit]	adj. 复杂的
collapsibility [kəlæpsi'biliti]	n. 退让性
contraction [kən'trækʃən]	n. 收缩
ram [ræm]	vt. 夯实
pattern ['pætən]	n. 模型，木模
pneumatic [nju:'mætik]	adj. 气动的
jolt [dʒəult]	vi. 振动，摇动
sling [sliŋ]	v. 抛 (砂)
flask [flæsk]	n. 砂箱
cope [kəup]	n. 上砂箱
drag [dræg]	n. 下砂箱
sprinkle ['spriŋkl]	vt. 撒
epoxy [e'pɔksi]	n. 环氧树脂 (胶)
taper ['teipə]	n. 锥度，起模斜度
draft [dræft]	n. 拨模斜度
foamed [fəumd]	adj. 泡沫的
polystyrene [,pɔli'stairi:n]	n. 聚苯乙烯
sprue [spru:]	n. 直浇口
runner ['rʌnə]	n. 内浇口，横浇口
basin ['beisən]	n. 浇口杯
deficient [di'fiʃənt]	adj. 不足的，缺乏的
shrinkage ['ʃriŋkidʒ]	n. 收缩
subtle ['sʌtl]	adj. 细微的
porosity [pə:'rɔsiti]	n. 多孔，缩松
porous ['pɔ:rəs]	adj. 多孔的
void [vɔid]	n. 空间
integrally ['intigrəli]	adv. 整体地
organic [ɔ:'gænɪk]	adj. 有机的
bonding ['bɔndiŋ]	n. 黏结剂

sodium ['səʊdjəm]	n. 钠
silicate ['silikeit]	n. 硅酸盐
plug [plʌg]	n. 塞
ladle ['leidl]	n. 浇勺，铁水包
alumina [ə'ljuːminə]	n. 氧化铝
line [lain]	v. 做内衬
susceptible [səs'septibl]	adj. 容易的
columnar [kə'lʌmnə]	adj. 柱状的
dendritic [den'dritik]	adj. 树枝的
burnout ['bəːnəut]	n. 熔蚀
risering ['raizəriŋ]	n. 冒口
protrusion [prə'truʒən]	n. 凸出物
fin [fin]	n. 周缘翅边
boss [bɔs]	n. 表面凸出部分
snag [snæg]	n. 毛刺，凸出物；vt. 清除（毛刺，浇口等）
chisel ['tʃizl]	n. 凿子，凿刀
chipping ['tʃipiŋ]	n. 修整，清理
be applicable to (sb/sth)	适用于
be referred to as	被称为
gross shrinkage	缩孔
make up for	补偿
be put in place	放置在该放的位置上
be susceptible to	易于
be resistant to	对……有抵抗力

⇨ Notes

[1] Green sand generally consists of silica sand and additives coated by rubbing the sand grains together with clay uniformly wetted with water.

参考译文：型砂通常含有石英砂和添加剂，通过砂粒与用水均匀溅湿的黏土的搅拌，使砂粒及添加剂表面包履一层黏结薄膜。

(1) 过去分词短语 coated by...with water 作定语修饰前置词 silica sand and additives。

(2) 动名词短语 rubbing the sand...with water 是介词 by 的宾语。其中，uniformly wetted with water 是过去分词短语作定语修饰 clay。

[2] As a result, the last portion of the casting to freeze will be deficient in metal and, in the absence of a supplemental metal-feed source, will result in some form of shrinkage.

参考译文：结果，铸件最后凝固的部分会缺少金属，而且，在缺少补缩金属源的情况下，将会产生某种形式的缩孔。

(1) 本句主语是 the last portion，含有两个并列谓语 will be deficient 和 will result in。

(2) 不定式短语 to freeze 作定语修饰前置词 the casting。

(3) feed 译为 "补缩"。

[3] Cores are placed in molds wherever it is necessary to preserve the space it occupies in the mold as a void in the resulting castings.

参考译文：型芯可放在铸型中需要之处，占据空间，以便在最后的铸件中形成孔洞。

(1) wherever it is necessary 是定语从句修饰 molds。

(2) 不定式短语 to preserve the space…as a void 作目的状语，修饰主句 cores are placed in molds。

(3) it occupies in the mold 是定语从句修饰前置词加 the space。

[4] The pouring process must be carefully controlled, since the temperature of the melt greatly affects the degree of liquid contraction before solidification, the rate of solidification, which in turn affects the amount of columnar growth present at the mold wall, the extent and nature of the dendritic growth, the degree of alloy burnout, and the feeding characteristics of the risering system.

参考译文：浇注过程必须小心控制，因为熔化温度会大大影响凝固前液态金属收缩的程度和凝固速率，这又将影响铸型中针状物成长的数量、程度和树枝状成长物的性质，以及合金溶蚀的程度和冒口系统的补缩特性。

(1) 从 since 开始至最后是原因状语从句修饰前面的主句。其中的 affects 后接两个宾语 the degree 和 the rate。

(2) which 至句末是非限制性定语从句修饰先行词 the rate of solidification。其中的 affects 后接 5 个宾语：the amount, the extent, nature, the degree, the feeding characteristics。

⌧ Questions

1. What is the largest bronze statue in existence today?
2. What basic factors should be depended on when selecting the best bronze to produce a particular part?
3. What is green sand molding used to produce?
4. What are the defects of wood patterns of sand casting?
5. Why must the pouring process be carefully controlled?

PART 3 Mechanical Elements

Unit 11 Gears

Spur and Helical Gears

A gear having tooth elements that are straight and parallel to its axis is known as a spur gear. A spur pair can be used to connect parallel shafts only. Parallel shafts, however, can also be connected by gears of another type, and a spur gear can be mated with a gear of a different type(Figure11-1).

Figure11-1 Spur gears

To prevent jamming as a result of thermal expansion, to aid lubrication, and to compensate for unavoidable inaccuracies in manufacture, all power-transmitting gears must have backlash. This means that on the pitch circles of a mating pair, the space width on the pinion must be slightly greater than the tooth thickness on the gear, and vice versa. On instrument gears, backlash can be eliminated by using a gear split down its middle; one half being rotatable relative to the other [1]. A spring forces the split gear teeth to occupy the full width of the pinion space.

Helical gears have certain advantages; for example, when connecting parallel shafts they have a higher load-carrying capacity than spur gears with the same tooth numbers and cut with the same cutter. Because of the overlapping action of the teeth, they are smoother in action and can operate at higher pitch-line velocities than spur gears. The pitch-line velocity is the velocity of the pitch circle. Since the teeth are inclined to the axis of rotation, helical gears create an axial thrust. If used singly, this thrust must be absorbed in the shaft bearings. The thrust problem can be overcome by cutting two sets of opposed helical teeth on the same blank. Depending on the method of manufacture, the gear may be of the continuous-tooth herringbone variety or a double-helical gear with a space between the two halves to permit the cutting tool to run out. Double-helical gear is well suited for the efficient transmission of power at high speeds.

Helical gears can also be used to connect nonparallel, non-intersecting shafts at any angle to one another. Ninety degrees is the commonest angle at which such gears are used.

Worm and Bevel Gears

In order to achieve line contact and improve the load carrying-capacity of the crossed-axis helical gears, the gear can be made to curve partially around the pinion, in somewhat the same that a nut envelops a screw [2]. The result would be a cylindrical worm and gear.

Worm gears provide the simplest means of obtaining large ratios in a single pair. They are usually less efficient than parallel-shaft gears, however, because of an additional sliding moving along the teeth. Because of their similarity, the efficiency of a worm and gear depends on the same factors as the efficiency of a screw. Single-thread worms of large diameter have small lead angles and low efficiencies. Multiplied-thread worms have larger lead angles and higher efficiencies (Figure 11-2).

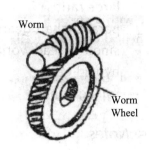

For transmitting rotary motion and torque around corners, bevel gears are commonly used. The connected shafts, whose axes would intersect if extended, are usually but not necessarily at right angles to one another.

Figure 11-2 Worm gear set

When adapted for shafts that do not intersect, spiral bevel gears are called hypoid gears. The pitch surfaces of these gears are not rolling cones, and the ratio of their mean diameters is not equal to the speed ratio. Consequently, the pinion may have few teeth and be made as large as necessary to carry the load.

The profiles of the teeth on bevel gears are not involutes; they are of such a shape that the tools for cutting the teeth are easier to make and maintain than involutes cutting tools. Since bevel gears come in pairs, as long as they are conjugate to one another they need not be conjugate to other gears with different tooth numbers.

◇ New Words and Expressions

lubrication [ˌluːbrɪˈkeɪʃən]	n.	润滑
backlash [ˈbæklæʃ]	n.	侧向间歇
blank [blæŋk]	n.	配件
nut [nʌt]	n.	螺母
screw [skruː]	n.	螺钉，丝杠
involute [ˈɪnvəluːt]	n.	渐开线
spur gear		直齿轮
thermal expansion		热膨胀
the pitch circles		节圆
space width		间歇宽度
instrument gear		仪表齿轮
the split gear		拼合齿轮
pitch-line velocities		节线速度
an axial thrust		轴线推力
double-helical gears		人字齿轮
non-intersecting		不相交
worm gears		蜗轮蜗杆

crossed-axis	交叉轴
large ratios	大速比
parallel-shaft gears	平行轴齿轮
single-thread worms	单线蜗杆
hypoid gear	准双曲面齿轮
rolling cones	滚锥

⇨ Notes

[1] On instrument gears, backlash can be eliminated by using a gear split down its middle, one half being rotatable relative to the other.

参考译文：在仪表齿轮上可以利用分开的拼合齿轮来消除侧向间隙，它的一半可相对于另一半转动。

其中 one half being rotatable... 为复合结构，表示伴随情况。

[2] In order to achieve line contact and improve the load carrying capacity of the crossed-axis helical gears, the gear can be made to curve partially around the pinion, in somewhat the same way that a nut envelops a screw.

参考译文：为了使交叉轴斜齿轮获得线接触并改进其承载能力，可以把大齿轮做成弯曲的部分抱在小齿轮上，有点像螺母套在螺钉上一样。

⊠ Questions

1. What is the function of a gear?
2. What are the features of a helical gear?
3. What are the features of a spur gear?

Reading Material: Spur Gears

Gears, defined as toothed members transmitting rotary motion from one shaft to another, are among the oldest devices and inventions of man. In about 2600 B.C., the Chinese are known to have used a chariot incorporating a complex series of gears. Aristotle, in the fourth century B.C., wrote of gears as if they were commonplace. In the fifteenth century A.D., Leonardo da Vinci designed a multitude of devices incorporating many kinds of gears.

Among the various means of mechanical power transmission (including primarily gears, belts, and chains), gears are generally the most rugged and durable [1]. Their power transmission efficiency is as high as 98 percent. On the other hand, gears are usually more costly than chains and belts. As would be expected, gear manufacturing cost increase sharply with increased precision—as required for the combination of high speeds and heavy loads, and for low noise levels[2]. (Standard tolerances for various degrees of manufacturing precision have been established by the AGMA: American Gear Manufacturers Association.)

Spur gears are the simplest and most common type of gears. They are used to transfer motion between parallel shafts and have teeth that are parallel to the shaft axes.

The basic requirement of gear-tooth geometry is the provision of angular velocity ratios that are exactly constant. For example, the angular velocity ratio between a 20-tooth and a 40-tooth gear must be precisely 2 in every position. (It must not be, for example, 1.99 as a given pair of teeth comes into mesh, and then 2.01 as they go out of mesh.) Of course, manufacturing inaccuracies and tooth deflections will cause slight deviations in velocity ratio, but acceptable tooth profiles are based on theoretical curves that meet this criterion.

As gears mesh, both rolling and sliding occur, causing pitting and wear; pitting is primarily a consequence of rolling contact and wear is the result of sliding contact[3]. Pitting is initiated when contact stresses are high; it is hastened by sliding action. Although sliding causes wear, it also creates hydrodynamic action that counteracts wear. This relative motion, minute as it may be, is sufficient to provide conditions necessary for hydrodynamic action: a narrow, wedge-shaped opening between surfaces that have relative motion. Each time a gear goes through mesh with its accompanying rolling and sliding, surface and subsurface material is subjected to compressive, shear, and tensile stresses. The result is slow surface destruction in the form of pockmarks and metal erosion. These phenomena, for clarity, are often described as occurring separately, but this is not the case in practice. Two or more may occur simultaneously. In fact, one type may promote the other. The result is roughening of tooth surface, alteration of tooth profile, and loss of conjugate motion. Failure has definitely occurred when performance is unsatisfactory because of noise, vibration, or overheating.

Pitting first occurs at the pitch line, where the absence of sliding favors early breakdown of protective oil films. Excessive contact stress causing surface fatigue is the real cause of pitting. This may be due to (1) a narrow gear width; (2) radii of the involute surfaces being too small; (3) frequent overloads. The smaller the radius of either surface, the narrower is the contact band and the greater the unit stress. After what is usually a very large number of stress repetitions, surface failure may occur. Minute cracks form in and below the surface, then grow and join. Eventually, small bits of metal are separated and forced out, leaving pits.

The least expensive gear material is usually ordinary cast iron, ASTM (or AGMA) grade 20. Grades 30, 40, 50, and 60 are progressively stronger and more expensive. Cast iron gears typically have greater surface fatigue strength than bending fatigue strength. Their internal damping tends to make them quieter than steel gears. Nodular cast iron gears have substantially greater bending strength, together with good surface durability. A good combination is often a steel pinion mated to a cast iron gear.

Unheat-treated steel gears are relatively inexpensive, but have low surface endurance capacity. Heat-treated steel gears must be designed to resist warpage; hence, alloy steels and oil quenching are usually preferred. For hardnesses over 250-350 Bhn, machining must usually be done before hardening[4]. Greater profile accuracy is obtained if the surfaces are finished after heat-treating, as by grinding. But if grinding is done, care must be taken to avoid residual tensile stresses at the surface.

Of the non-ferrous metals, bronzes are most often used for making gears. Nonmetallic gears (nylon and other plastics) are generally quiet, durable, reasonably priced, and often can operate under light loads without lubrication[5]. Their teeth deflect more easily than those of corresponding metal gears. This promotes effective load sharing among teeth in simultaneous contact. Since nonmetallic materials have low thermal conductivity, special cooling provisions may be required. Furthermore, these materials have relatively high coefficients of thermal expansion, and this may require installation with greater backlash than metal gears.

Nonmetallic gears are usually mated with cast iron or steel pinions. For best wear resistance, the hardness of the mating metal pinion should be at least 300 Bhn. Design procedures for gears made of plastics are similar to those for gears made of metals, but are not yet as reliable. Hence, prototype testing is even more important than for metal gears.

◇ *New Words and Expressions*

gear [gɪə(r)]	n. 齿轮；v. 调整，(使) 适合，换挡
shaft [ʃɑ:ft]	n. 轴，杆状物
chariot ['tʃærɪət]	n. 战车，托车，托架
multitude ['mʌltɪtju:d]	n. 多数，群众
belt [belt]	n. 带子，地带
rugged ['rʌgɪd]	adj. 高低不平的，崎岖的，粗糙的，有皱纹的
durable ['djʊərəbl]	adj. 持久的，耐用的
mesh [meʃ]	vt. 啮合，编织；vi. 落网，相啮合
tolerances ['tɔlərənsiz]	n. 公差，宽容；vt. 给规定公差
parallel ['pærəlel]	adj. 平行的；n. 平行线；v. 相应，平行
pitting ['pɪtɪŋ]	n. 蚀损斑，氢气泡疤，点蚀
hasten ['heɪsn]	v. 催促，赶紧
hydrodynamic ['haɪdrəʊdaɪ'næmɪk]	adj. 水力的
counteract [ˌkaʊntər'ækt]	vt. 抵消，中和，阻碍
wedge-shaped ['wedʒʃeɪpt]	楔形
subsurface ['sʌb'sɜ:fɪs]	adj. 表面下的，地下的
compressive [kəm'presɪv]	adj. 有压缩力的
tensile ['tensaɪl]	adj. 可拉长的，可伸长的，[物] 张力的，拉力的
phenomena [fə'nɒmɪnə]	n. 现象 (复数)
roughening ['rʌfənɪŋ]	n. 粗加工，粗糙法
conjugate ['kɒndʒəgeɪt]	adj. 成对的，配合的，结合的，共轭的
vibration [vaɪ'breɪʃn]	n. 振动，颤动，摇动，摆动

pitch [pɪtʃ]	n. 螺距，分度
involute ['ɪnvəljuːt]	adj. 纷乱的，复杂的；n. [数] 渐开线
fatigue [fə'tiːg]	n. 疲乏，疲劳；v. 使疲劳，使心智衰弱
warpage ['wɔːpeɪdʒ]	n. & v. 翘曲，扭曲，热变形
coefficient [ˌkəʊɪ'fɪʃnt]	n. [数] 系数
pinion ['pɪnjən]	n. 小齿轮
prototype ['prəʊtətaɪp]	n. 原型，模型，典型，榜样

⇨ Notes

[1] Among the various means of mechanical power transmission (including primarily gears, belts, and chains), gears are generally the most rugged and durable.

参考译文：在众多的机械动力传动方式中(主要包括齿轮传动、传动带传动和链传动)，齿轮传动通常是最坚固和最耐用的。

[2] As would be expected, gear manufacturing costs increase sharply with increased precision—as required for the combination of high speeds and heavy loads, and for low noise levels.

参考译文：正如所料，高速重载及无噪音传动要求齿轮有高的精度，这使齿轮加工成本大大增加。

[3] As gears mesh, both rolling and sliding occur, causing pitting and wear; pitting is primarily a consequence of rolling contact and wear is the result of sliding contact.

参考译文：当齿轮啮合时既有滚动又有滑动，这样就引起点蚀和磨损。点蚀主要是由滚动接触造成的，而磨损是由滑动接触造成的。

[4] For hardnesses over 250-350 Bhn, machining must usually be done before hardening.

参考译文：当硬度大于250～350 Bhn 时，切削加工一般在淬火以前进行。

Bhn 为 Brinel hardness number [冶]布氏硬度值。

[5] Of the non-ferrous metals, bronzes are most often used for making gears. Nonmetallic gears (nylon and other plastics) are generally, quiet, durable, reasonably priced, and often can operate under light loads without lubrication.

参考译文：对于有色金属来说，青铜最常用来制造齿轮。非金属齿轮(尼龙和其他的塑料)通常运转无噪声、耐用、价格适中，一般用在低载不需要润滑的场合。

⊠ Questions

1. Please describe briefly the history of gears.
2. Why are gears generally regarded as the most rugged and durable means of mechanical power transmission?
3. What is the function of spur gears?
4. What is the real cause of pitting? Why?

Unit 12 Shafts and Couplings

Virtually all machines contain shafts. The most common shape for shafts is circular and the cross section can be either solid or hollow (hollow shafts can result in weight savings). Rectangular shafts are sometimes used, as in screw driver blades, socket wrenches and control knob stems.

A shaft must have adequate torsional strength to transmit torque and not be over stressed. It must also be torsionally stiff enough so that one mounted component does not deviate excessively from its original angular position relative to a second component mounted on the same shaft. Generally speaking, the angle of twist should not exceed one degree in a shaft length equal to 20 diameters.

Shafts are mounted on bearings and transmit power through such devices as gears, pulleys, cams and clutches. These devices introduce forces which attempt to bend the shaft; hence, the shaft must be rigid enough to prevent overloading of the supporting bearings. In general, the bending deflection of a shaft should not exceed 0.01 inches per ft of length between bearing supports.

In addition, the shaft must be able to sustain a combination of bending and torsional loads. Thus an equivalent load must be considered which takes into account both torsion and bending, Also, the allowable stress must contain a factor of safety which includes fatigue, since torsional and bending stress reversals occur.

For diameters less than 3 in., the usual shaft material is cold-rolled steel containing about 0.4 percent carbon. Shafts are either cold-rolled or forged in sizes from 3 in. to 5 in. For sizes above 5 in., shafts are forged and machined to size [1]. Plastic shafts are widely used for light load applications. One advantage of using plastic is safety in electrical applications, since plastic is a poor conductor of electricity.

Components such as gears and pulleys are mounted on shafts by means of key. The design of the key and the corresponding keyway in the shafts must be properly evaluated. For example, stress concentrations occur in shafts due to keyways, and the material removed to form the keyway further weakens the shaft.

If shaft are run at critical speeds, severe vibrations can occur which can seriously damage a machine. It is important to know the magnitude of these critical speeds so that they can be avoided. As a general rule of thumb, the difference between the operating speed and the critical speed should be at least 20 percent[2].

Many shafts are supported by three or more bearings, which means that the problem is statically indeterminate. Texts on strength of materials give methods of solving such problems. The design effort should be in keeping with the economics of a given situation. For example, if one line shaft supported by three or more bearings is needed, it probably would be cheaper to make conservative assumptions as to moments and design it as though it were determinate. The extra cost of an oversize shaft may be less than the extra cost of an elaborate design analysis.

Another important aspect of shaft design is the method of directly connecting one shaft to another. This is accomplished by devices such as rigid and flexible couplings.

A coupling is a device for connecting the ends of adjacent shafts. In machine construction, couplings are used to effect a semipermanent connection between adjacent rotating shafts. The connection is permanent in the sense that it is not meant to be broken during the useful life of the machine, but it can be broken and restored in an emergency or when worn parts are replaced.

There are several types of shaft couplings, their characteristics depend on the purpose for which they are used. If an exceptionally long shaft is required in a manufacturing plant or a propeller shaft on a ship, it is made in sections that are coupled together with rigid couplings. A common type of rigid coupling consists of two mating radial flanges (disks) that are attached by key-driven hubs to the ends of adjacent shaft sections and bolted together through the flanges to form a rigid connection. Alignment of the connected shafts is usually effected by means of a rabbet joint on the face of the flanges [3].

In connecting shafts belonging to separate devices (such as an electric motor and a gearbox), precise aligning of the shafts is difficult and a flexible coupling is used. This coupling connects the shafts in such a way as to minimize the harmful effects of shaft misalignment[4]. Flexible couplings also permit the shafts to deflect under their separate systems of loads and to move freely (float) in the axial direction without interfering with one another. Flexible couplings can also serve to reduce the intensity of shock loads and vibrations transmitted from one shaft to another.

◇ New Words and Expressions

coupling ['kʌplɪŋ]	n. 耦合，结合，联结，连接
rectangular [rek'tæŋgjʊlə]	adj. 矩形的，成直角的
cross section	截面，横断面，剖面
screw [skruː]	n. 螺旋，螺丝钉
screw driver	螺丝刀，螺丝起子，改锥
blade [bleɪd]	n. 叶片，桨片
socket ['sɒkɪt]	n. 插座，插口，套筒扳手
wrench [rentʃ]	n. 扳手，扳钳，猛扭
socket wrench	套筒扳手
knob [nɒb]	n. 节，球块，旋钮，圆形把手
stem [stem]	n. 杆，棒，柱，轴

torsional ['tɔːʃənəl]	adj. 扭转的，扭力的
mounted ['maʊntɪd]	adj. 安装好的，固定好的
deviate ['diːvɪeɪt]	v. 脱离，使偏离
deviate from	背离，偏离，与……有偏差
twist [twɪst]	v. 使扭转，扭，使转动
clutch [klʌtʃ]	n. 离合器；vt. 抓住
bending ['bendɪŋ]	n. 弯曲度；v. 弯曲
deflection [dɪ'flekʃ(ə)n]	n. 弯曲，弯曲度，挠度
reversal [rɪ'vɜːs(ə)l]	n. 颠倒，相反，反向，改变方向，倒转
cold-roll	n. & v. 冷轧，冷轧机
forge [fɔːdʒ]	n. & v. 锻造，打制，锻工车间
key [kiː]	n. 键
keyway ['kiːweɪ]	n. [机]键槽，锁槽，销座
indeterminate [ˌɪndɪ'tɜːmɪnət]	adj. 不确定的，模糊的，含混的
statically indeterminate	静不定，超静定
conservative [kən'sɜːvətɪv]	n. 保守派，有裕量
adjacent [ə'dʒeɪs(ə)nt]	adj. 邻近的
semipermanent [ˌsemɪ'pɜːmənənt]	adj. 非永久性的，暂时的
propeller [prə'pelə]	n. [航][船] 螺旋桨，推进器
flange [flæn(d)ʒ]	n. 凸缘，法兰
hub [hʌb]	n. 中心部分，衬套，轮毂
bolt [bəʊlt]	n. 螺栓，螺钉
alignment [ə'laɪnm(ə)nt]	n. 直线对准，调准，对中心
rabbet ['ræbɪt]	n. 插孔，塞孔，缺口
gearbox ['gɪəbɒks]	n. 变速箱，齿轮箱
flexible ['fleksɪb(ə)l]	adj. 灵活的，柔韧的，易弯曲的
shock [ʃɒk]	n. 冲击，冲撞，打击

⇨ *Notes*

[1] Shafts are either cold-rolled or forged in sizes from 3 in. to 5 in. For sizes above 5 in., shafts are forged and machined to size.

参考译文：当轴的直径尺寸大于 5 英寸时，则采用锻造毛坯，然后经过机械加工达到要求的尺寸。

machined 这里的意思为"经过机械加工"。

[2] It is important to know the magnitude of these critical speeds so that they can be avoided. As a general rule of thumb, the difference between the operating speed and the critical speed should be at least 20 percent.

参考译文：重要的是要知道这些临界速度的大小，使它们可以避免。作为一个通用的经验法则，操作速度和临界速度之间的差异应至少为百分之二十。

rule of thumb 意为"经验法则"。

[3]　Alignment of the connected shafts is usually effected by means of a rabbet joint on the face of the flanges.

参考译文：被连接的两根轴之间的找正通常是借助于法兰盘面上的企口接合来实现的。

by means of 意为"依靠，借助于"。

[4]　This coupling connects the shafts in such a way as to minimize the harmful effects of shaft misalignment.

参考译文：这种联轴器以一种能够把轴线不重合所造成的有害影响减少至最低的方式将两根轴连接到一起。

in such a way as 意为"以这样一种方式"。

Questions

1. What features should a shaft have?
2. What is the function of shafts?
3. What is a coupling and what is its function?
4. In what way are gears and pulleys mounted on shafts?

Reading Material: Couplings

Couplings

A coupling is a mechanical device for uniting or connecting parts of a mechanical system. This chapter concerns itself with those couplings which are used on machine shafts for the purpose of transmitting torque. Couplings may be employed for a permanent or semi-permanent connection between shafts or for disconnection of machine components to permit one member to run while the other is stationary.

Commercial shafts are limited in length by manufacturing and shipping requirements so that it is necessary to join sections of long transmission shafts with couplings. Couplings are also required to connect the shaft of a driving machine to a separately built driven unit. Permanent couplings are referred to simply as couplings, while those which may be readily engaged to transmit power, or disengaged when desired, usually are called clutches[1].

Rigid Couplings

Rigid couplings are permanent couplings which by virtue of their construction have essentially no degree of angular, axial, or rotational flexibility; they must be used with collinear shafts.

The flange coupling shown in Figure 12-1 is perhaps the most common coupling. It has the advantage of simplicity and low cost, but the connected shafts must be accurately aligned to prevent severe bending stresses and excessive wear in the bearings. The length of the hub is determined by the length of key required, and the hub diameter is approximately twice the bore.

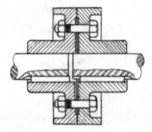

Figure 12-1   Flange coupling

The thickness of the flange is determined by the permissible bearing pressure on the bolts. Although usually not critical, the shearing stress on the cylindrical area where the flange joins the hub should be checked [2].

When large flanges are objectionable, compression couplings similar to the coupling in Figure12-2 may be used. Torque is transmitted by keys between the shafts and the cones and by friction between the cones and the outer sleeve [3]. The inner cones may be split so that they will grip the shaft when drawn together.

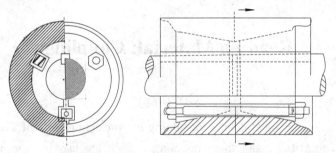

Figure 12-2   Compression coupling

A collar coupling, shown in Figure12-3, consists of a cylindrical collar pressed over the ends of the two collinear shafts being connected, approximately one-half of the collar connecting each shaft. Usually one or more radial pins completely through each shaft and the collar, or setscrews, may be used to ensure that there is no undesired radial movement. A variation of the collar coupling results if a cylindrical part of one shaft fits concentrically into a female bored portion in the second shaft and pins, setscrews, or other suitable fasteners are used [4].

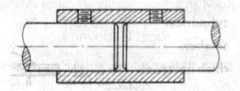

Figure 12-3   Collar coupling with set screws

Flexible Couplings

Upon occasion it is desired that couplings be able to accommodate reasonable amounts of axial angularity between shafts, a small amount of eccentricity between parallel shafts, or axial movement of shafts during use. Flexible couplings may be employed for any or all of these cases. In addition, torsionally flexible couplings may be employed to absorb some of the torque in the shaft, or to permit large amounts of torsional flexibility as in fluid couplings or torque converters.

Perhaps the simplest flexible coupling permitting small amounts of axial misalignment and/or torsional flexibility consists of a piece of elastic material (such as neoprene, tflon, or similar material) bonded to two separated collinear cylindrical members which in turn can be fastened onto the two shafts by means of setscrews or other fasteners [5].

Such an elastic-material bounded coupling is shown in Figure 12-4.

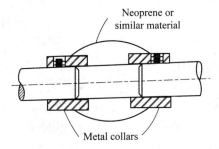

Figure 12-4 Simple elastic material bonded coupling

◇ New Words and Expressions

coupling ['kʌplɪŋ]	n. 联轴器,连接器
device [dɪ'vaɪs]	n. 装置,器械,机构
transmit [trænz'mɪt]	vt. 传递,转动
torque [tɔːk]	n. 转矩,扭矩
engage [ɪn'geɪdʒ]	vt. 连接,啮合
disengage [ˌdɪsɪn'geɪdʒ]	vt. 脱离
clutch [klʌtʃ]	n. 离合器
rigid ['rɪdʒɪd]	adj. 刚性的,刚硬的
angular ['æŋgjʊlə]	adj. 角度的
axial ['æksɪəl]	adj. 轴向的,轴的,成轴的
collinear [kə'lɪnɪə]	adj. 共线的,在同一线上的
flange [flæn(d)ʒ]	n. 法兰,凸缘
align [ə'laɪn]	vt. 使成一线(行)
hub [hʌb]	n. 毂,中心
permissible [pə'mɪsɪb(ə)l]	adj. 容许的

bolt [bəʊlt]	n. 螺栓
cylindrical [sɪ'lɪndrɪkəl]	adj. 圆柱体的，圆筒形的
objectionable [əb'dʒekʃ(ə)nəb(ə)l]	adj. 不能采用的，不适合的
cone [kəʊn]	n. 锥面，椎体，锥形物
sleeve [sliːv]	n. 套筒
grip [grɪp]	vt. & n. 紧夹，紧握
collar ['kɒlə]	n. 套管，轴环
setscrew ['setskruː]	n. 定位螺钉，固定螺钉
variation [veərɪ'eɪʃ(ə)n]	n. 变化，变异
concentrically [kən'sentrɪklɪ]	adv. 同轴，同心
female ['fiːmeɪl]	adj. 内孔的，凹形的
bore [bɔː]	v. 镗孔
accommodate [ə'kɒmədeɪt]	v. 容纳，适应
angularity [ˌæŋgjʊ'lærətɪ]	n. 斜倾度
eccentricity [ˌeksen'trɪsɪtɪ]	n. 偏心度，偏心距
absorb [əb'zɔːb]	vt. 缓冲，吸收，减震
converter [kən'vɜːtə]	n. 变换器，转换器
misalignment [mɪsə'laɪnmənt]	n. 不对准，不同轴
neoprene ['niːə(ʊ)priːn]	n. 氯丁橡胶
teflon ['teflɔn]	n. 聚四氟乙烯
concern oneself with	涉及，关心
for the purpose of	为了，以便
by virtue of	由于，因为，凭借
upon occasion	有时，间或
in turn	依次，按顺序，一个一个的

⇨ Notes

[1] Permanent couplings are referred to simply as couplings, while those which May be readily engaged to transmit power, or disengaged when desired, usually are called clutches.

参考译文：固定式联轴器简称为联轴器，而那些随时可以连接起来传递动力并在需要时可以脱开的联轴器常称为离合器。

which may be…when desired 为一定语从句，修饰先行词 those。在这个定语从句中，when desired 的完整结构是 when they are desired，作从句中的谓语动词 disengaged 的状语；另外，disengaged 前省略了已出现的情态动词和助动词 may be。

[2] Although usually not critical, the shearing stress on the cylindrical area where the flange joins the hub should be checked.

参考译文：在法兰与毂连接的圆柱面上的剪应力虽然通常不很重要，但也应进行校核。

句中 where...the hub 为一定语从句，修饰先行词 area，关系副词在定语从句中作地点状语。

[3] Torque is transmitted by keys between the shafts and the cones and by friction between the cones and the outer sleeve.

参考译文：扭矩由轴与圆锥体之间的键以及圆锥体与外套之间的摩擦来传递。

该句是一个被动句，谓语动词所表示的动作执行者有两个，一个是 keys，另一个是 friction。介词短语 between the shafts and the cones 作 keys 的定语，另一个介词短语 between the cones and the outer sleeve 作 friction 的定语。

[4] A variation of the collar coupling results if a cylindrical part of one shaft fits concentrically into a female bored portion in the second shaft and pins, setscrews, or other suitable fasteners are used.

参考译文：如果一根轴的圆柱部分同轴地套入另一根轴的内孔腔内，并用销子、定位螺钉或其它固定件加以固定，就构成了一种变化了的套管式联轴器。

该句为主从复合句，主句是 A variation ... results if 引导的是一个条件状语从句，这个从句又是一个并列句，第一个分句是 a cylindrical part...in the second shaft 句中谓语动词是 fits；第二个分句是 pins, setscrews, ... are used 是个被动句。两个分句是贯连的，即必须同时满足这两个条件，主句所述的结论才能成立。

[5] Perhaps the simplest flexible coupling permitting small amounts of axial misalignment and/or torsional flexibility consists of a piece of elastic material (such as neoprene, Teflon, or similar material) bonded to two separated collinear cylindrical members which in turn can be fastened onto the two shafts by means of setscrews or other fasterners .

参考译文：也许，允许少量轴向对中偏差和(或)扭转弹性的最简单的柔性联轴器具有如下结构：将一个弹性材料固接在两个分离共线的圆筒形结构件上，再通过定位螺钉或其它紧固件将圆筒形结构件固定在两个轴上。

分词短语 permitting ... torsional flexibility 作主语从句 flexible coupling 的定语。主句的谓语是 consists of，宾语是 a piece of elastic material。过去分词短语 bound ... members 作定语修饰 material。Which ... fasteners 为定语从句，修饰先行词 members, which 在定语从句中作主语。

⊠ Questions

1. What is a coupling?
2. What is a clutch?
3. Describe the main structure of collar couplings?
4. When is a flexible coupling needed?
5. Which characters do the rigid coupling have?

Unit 13 Clutches

A clutch is a device for quickly and easily connecting or disconnecting a rotatable shaft and rotating coaxial shaft.

Friction Clutches

Friction clutches have pairs of mating conical, disk, or ring-shaped surfaces and means for pressing the surfaces together. The pressure may be created by a spring or by a series of levers locked in position by the wedging action of a conical spool [1]. On a spring-loaded clutch the operator, by controlling the rate at which the spring pressure is applied to the clutch, can regulate the speed of clutch engagement and the torque applied to the driven shaft. There is always some slippage, however, and the efficiency of a friction clutch can never exceed 50 percent; i.e., during a clutching operation at least one-half of the input energy is lost by friction in the clutch and produces heat.

The friction surfaces on clutches should have a high coefficient of friction be able to conduct the heat away rapidly. These properties are difficult to obtain in a single material and for this reason, one of each pair of mating surface is usually metallic, while the other is either leather, cork, or an asbestos-based facing riveted to a metal plate. Some friction clutches are run dry, while others operate in oil. Dry clutches have a higher coefficient of friction than wet clutches, but the oil helps to carry off the heat.

Figure13-1 shows a half-section of a multiple-plate disk clutch in which input member 2 is keyed to the driving shaft 1 and output member 3 is keyed to the driven shaft 4. The friction plate b has external gear teeth or splines that mate with teeth on the inside of member 2, while friction plate c has internal teeth that mate with external teeth on member 3. Plate b can slide axially in 2, while plate c can slide axially on 3. The clutch is engaged by moving the spool to the left, which, by a wedging action, rotates the lever about the pivot p and creates a force that squeezes the plates together.

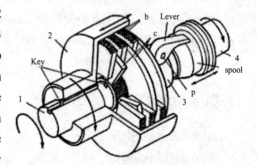

Figure 13-1 Multiple-plate disk clutch

Magnetic Clutches

Magnetic-particle clutches have an annular (ring-shaped) gap between the driving and driven members that is filled with powdered iron and graphite. When a magnetic field is induced across the gap by a direct-current control coil, the iron particle form chains across the gap and transmit a torque that depends on the strength of the field [2]. Controlled by varying the current, the load can be engaged smoothly and there is no slippage when the clutch is transmitting the torque for which it was designed [3].

◇ New Words and Expressions

coaxial [kəʊ'æksɪəl]	adj. 同轴的
spring [sprɪŋ]	n. 弹簧
facing ['fesɪŋ]	n. 衬片
splines [sp'lɪnz]	n. 花键
pivot ['pɪvət]	n. 枢轴
slippage ['slɪpɪdʒ]	n. 滑动
ring-shaped [rɪŋ'ʃept]	adj. 环形的
graphite ['græfˌaɪt]	n. 石墨
a rotatable shaft (driven shaft)	从动轴
a rotating shaft (driving shaft)	主动轴
friction clutches	摩擦离合器
a conical spool	锥形短管轴
coefficient of friction	摩擦系数
dry clutches	干式离合器
a multiple-plate disk clutch	多片式圆盘离合器
wet clutches	湿式离合器
magnetic clutches	电磁离合器
an annular gap	环状间隙
powdered iron	铁粉
direct-current control coil	直流控制线圈

⇨ Notes

[1] The pressure may be created by a spring or by a series of levers locked in position by the wedging action of a conical spool.

参考译文：此压力由弹簧或一组控制杆产生，这组控制杆通过锥形短管轴的楔入锁在位置上。

其中 locked in position... 为过去分词短语作定语，修饰 levers。

[2] When a magnetic field is induced across the gap by a direct-current control coil, the iron particle form chains across the gap and transmit a torque depends on the strength of the field.

参考译文：当直流控制线圈越过间歇感应一磁场时，铁粉越过间隙形成许多链，并传送一扭矩，该扭矩大小取决于磁场强度。

其中 when a magnetic field… control coil 为时间状语从句；that depends on…为定语从句，修饰 a torque；主句中，the iron particles 为主语，form 和 transmit 为两个并列谓语。

[3] Controlled by varying the current, the load can be engaged smoothly and there is no slip page when the clutch is transmitting the torque for which it was designed.

参考译文：通过控制电流的变化，负载可平稳接合，而且在离合器传递预定的扭矩时不会滑动。

其中 Controlled by…current 为分词短语作状语；for which it…为定语从句，修饰 the torque。

⊠ Questions

1. What is the function of a clutch?
2. What are the components of a clutch?
3. How can we classify clutches?

Reading Material: Shafts, Clutches and Brakes

A shaft is a rotating or stationary member, usually of circular cross section, having mounted upon it such elements as gears, pulleys, flywheels, cranks, sprockets, and other power-transmission elements. Shafts may be subjected to bending, tension, compression, or torsional loads, acting singly or in combination with one another. When they are combined, one may expect to find both static and fatigue strength to be important design considerations, since a single shaft may be subjected to static stresses, completely reversed, and repeated stresses, all acting at the same time.

The word "shaft" covers numerous variations, such as axles and spindles. An axle is a shaft, either stationary or rotating, not subjected to torsion load. A short rotating shaft is often called a spindle.

When either the lateral or the torsional deflection of a shaft must be held to close limits, the shaft must be sized on the basis of deflection before analyzing the stresses[1]. The reason for this is that, if the shaft is made stiff enough so that the deflection is not too large, it is probable that the resulting stresses will be safe. But by no means should the designer assume that they are safe; it is almost always necessary to calculate them so that he knows they are within acceptable limits. Whenever possible, the power-transmission elements, such as gears or pulleys, should be located close to the supporting bearings. This reduces the bending moment and hence the deflection and bending stress.

Although the von Mises-Hencky-Goodman method is difficult to use in design of shaft, it probably comes closest to predicting actual failure. Thus it is a good way of checking a shaft that has already been designed or if discovering why a particular shaft has failed in service. Furthermore, there are a considerable number of shaft-design problems in which the dimensions are pretty well limited by other considerations, such as rigidity, and it is necessary for the designer to discover something about the fillet sizes, heat-treatment, and surface finish and whether or not shot peening is necessary in order to achieve the required life and reliability.

Because of the similarity of their functions, clutches and brakes are treated together in this lesson. In a simplified dynamic representation of a friction clutch, or brake, two inertias I_1 and I_2 traveling at the respective angular velocities ω_1 and ω_2, one of which may be zero in the case of brake, are to be brought to the same speed by engaging the clutch or brake. Slippage occurs because the two elements are running at different speeds and energy is dissipated during actuation, resulting in a temperature rise. In analyzing the performance of these devices we shall be interested in the actuating force, the torque transmitted, the energy loss and the temperature rise. The torque transmitted is related to the actuating force, the coefficient of friction, and the geometry of clutch or brake. This is a problem in statics, which will have to be studied separately for each geometric configuration. However, temperature rise is related to energy loss and can be studied without regard to the type of brake or clutch because the geometry of interest is the heat-dissipating surfaces. The various types of clutches and brakes may be classified as follows:

(1) Rim type with internally expanding shoes;

(2) Rim type with externally contracting shoes;

(3) Band type;

(4) Disk or axial type;

(5) Cone type;

(6) Miscellaneous type.

The analysis of all types of friction clutches and brakes uses the same general procedure. The following steps are necessary:

(1) Assume or determine the distribution of pressure on the frictional surfaces;

(2) Find a relation between the maximum pressure and the pressure at any point;

(3) Apply the condition of statical equilibrium to find (a) the actuating force, (b) the torque, and (c) the support reactions.

Miscellaneous clutches include several types, such as the positive-contact clutches, overload-release clutches, overrunning clutches, magnetic fluid clutches, and others.

A positive-contact clutch consists of a shift lever and two jaws. The greatest differences between the various types of positive clutches are concerned with the design of the jaws. To provide a longer period of time for shift action during engagement, the jaws may be ratchet-shaped, spiral-shaped, or gear-tooth-shaped. Sometimes a great many teeth or jaws are used, and they may be cut either circumferentially, so that they engage by cylindrical mating, or on the faces of the mating elements.

Although positive clutches are not used to the extent of the frictional-contact type, they do have important applications where synchronous operation is required.

Devices such as linear drives or motor-operated screw drivers must run to definite limit and then come to a stop. An overload-release type of clutch is required for these applications. These clutches are usually spring-loaded so as to release at a predetermined toque. The clicking sound which is heard when the overload point is reached is considered to be a desirable signal.

An overrunning clutch or coupling permits the driven member of a machine to "freewheel" or "overrun" because the driver is stopped or because another source of power increases the speed of the driven. This type of clutch usually uses rollers or balls mounted between an outer sleeve and an inner member having flats machined around the periphery[2]. Driving action is obtained by wedging the rollers between the sleeve and the flats. The clutch is therefore equivalent to a pawl and ratchet with an infinite number of teeth.

Magnetic fluid clutch or brake is a relatively new development which has two parallel magnetic plates. Between these plates is a lubricated magnetic powder mixture. An electromagnetic coil is inserted somewhere in the magnetic circuit. By varying the excitation to this coil, the shearing strength of the magnetic fluid mixture may be accurately controlled. Thus any condition from a full slip to a frozen lockup may be obtained.

◇ *New Words and Expressions*

shaft [ʃɑːft]	n. 轴
fillet ['fɪlɪt]	n. 圆角，倒角
peening ['piːnɪŋ]	v. 喷射 (加工硬化法)
brake [breik]	n. 制动器
stiff [stif]	adj. 刚硬的，刚的
flywheel ['flaiwiːl]	n. 飞轮
slippage ['slipidʒ]	n. 滑动
actuation [ˌæktjuˈeiʃən]	n. 驱动，开动
torsional ['tɔːʃənəl]	adj. 扭转的
coefficient [ˌkəuiˈfiʃənt]	n. 系数
axle ['æksl]	n. 心轴，轮轴
rim [rim]	n. 边缘，轮缘
spindle ['spindl]	n. 心轴，主轴
shoe [ʃuː]	n. 闸瓦，制动片 (块)
band [bænd]	n. 带，条
cone ['kəun]	n. 圆锥
miscellaneous [misəˈleiniəs]	adj. (混) 合的，杂项的
clicking [klikiŋ]	n. "咔嗒"声
assume [əˈsjuːm]	v. 假设，承担
freewheel ['friːˈwiːl]	n. 空转
statical ['stætikl]	adj. 静态的
coupling ['kʌpliŋ]	n. 联轴器
equilibrium [ˌiːkwiˈlibriəm]	n. 平衡

sleeve [sliːv]	n. 套筒
reaction [riːˈækʃən]	n. 反应，反力
flat [flæt]	n. 平面 (部分)；adj. 平的
periphery [pəˈrifəri]	n. 圆周，周边
wedge [wedʒ]	n. 楔形物
pawl [pɔːl]	n. 棘爪
powder [ˈpaudə]	n. 粉末
lever [ˈliːvə]	n. 杆，手柄
jaw [dʒɔː]	n. 夹抓
electromagnetic [iˈlektrəuˈmægnetik]	adj. 电磁的
ratchet [ˈrætʃit]	n. 棘轮
coil [kɔil]	n. 线圈
circumferentially [səkʌmfəˈrenʃəli]	adv. 周围地，圆周地
excitation [ˌeksiˈteiʃən]	n. 刺激，激励
lockup [ˈlɔkʌp]	n. 锁住
shot peening	喷丸硬化
bending moment	弯曲力矩
overload-release clutch	过载释放 (保护) 离合器
overrunning clutch	过速离合器
magnetic fluid clutch	磁液离合器

⇨ Notes

[1] When either the lateral or the torsional deflection of a shaft must be held to close limits, the shaft must be sized on the basis of deflection…

参考译文：如果轴的弯曲变形或扭转变形的限度非常严格，应该先根据变性条件确定轴的尺寸……。

这里 size 是动词，表示依一定的尺寸制造。

[2] …an inner member having flats machined around the periphery.

参考译文：……一个周边加工出数个平面的内部零件。

⊠ Questions

1. How many types of clutches and brakes are there?
2. What elements mount upon a shaft?
3. What is a shaft?
4. What is the advantage of von Mises-Hencky-Goodman method?
5. What does a positive-contact clutch consist of?

Unit 14 Rolling Contact Bearings

The concern of a machine designer with ball and roller bearings is fivefold as follows: (a) life in relation to load; (b) stiffness, i.e. deflections under load; (c) friction; (d) wear; (e) noise. For moderate loads and speeds the correct selection of a standard bearing on the basis of load rating will usually secure satisfactory performance. The deflection of bearing elements will become important where loads are high, although this is usually of less magnitude than that of the shafts or other components associated with the bearing. Where speeds are high special cooling arrangements become necessary which may increase frictional drag. Wear is primarily associated with the introduction of contaminants, and sealing arrangements must be chosen with regard to the hostility of the environment.

Because the high quality and low price of ball and roller bearings depends on quantity production, the task of the machine designer becomes one of selection rather than design. Rolling-contact bearings are generally made with steel which is through-hardened[1] to about 900 HV, although in many mechanisms special races are not provided and the interacting surfaces are hardened to about 600 HV. It is not surprising that, owing to the high stresses involved, a predominant form of failure should be metal fatigue, and a good deal of work is currently in progress intended to improve the reliability of this type of bearing. Design can be based on accepted values of life and it is general practice in the bearing industry to define the load capacity of the bearing as that value below which 90 percent of a batch will exceed a life of one million revolutions.

Notwithstanding the fact that responsibility for the basic design of ball and roller bearings rests with the bearing manufacturer, the machine designer must form a correct appreciation of the duty to be performed by the bearing and be concerned not only with bearing selection but with the conditions for correct installation.

The fit of the bearing races onto the shaft or onto the housings is of critical importance because of their combined effect on the internal clearance of the bearing as well as preserving the desired degree of interference fit. Inadequate interference can induce serious trouble from fretting corrosion[2]. The inner race is frequently located axially by abutting against a shoulder. A radius at this point is essential for the avoidance of stress concentration and ball races are provided with a radius or chamfer to allow space for this.

Where life is not the determining factor in design, it is usual to determine maximum loading by the amount to which a bearing will deflect under load. Thus the concept of "static load-carrying capacity"[3] is understood to mean the load that can be applied to a bearing, which is either stationary or subject to slight swiveling motions, without impairing its running qualities for subsequent rotational motion. This has been determined by practical experience as the load which when applied to a bearing results in a total deformation of the rolling element and raceway at any point of contact not exceeding 0.01 percent of the rolling-element diameter. This would correspond to a permanent deformation of 0.0025mm for a ball 25mm in diameter.

The successful functioning of many bearings depends upon providing them with adequate protection against their environment, and in some circumstances the environment must be protected from lubricants or products of deterioration of the bearing surfaces. Achievement of the correct functioning of seals is an essential part of bearing design. Moreover, seals which are applied to moving parts for any purpose are of interest to tribologists because they are components of bearing system and can only be designed satisfactorily on the basis of the appropriate bearing theory.

Notwithstanding their importance, the amount of research effort that has been devoted to the understanding of the behavior of seals has been small when compared with that devoted to other aspects of bearing technology.

◇ New Words and Expressions

fivefold ['faɪvfoʊld]	adj. 五倍的，五重的
stiffness ['stɪfnɪs]	n. 刚度，刚性，稳定性
deflection [dɪ'flekʃ(ə)n]	n. 偏向，挠曲，偏差
frictional ['frɪkʃənəl]	adj. 摩擦的，摩擦力的
drag [dræg]	n. 拖；拖累；vi. 拖曳；vt. 拖累；拖拉
wear [weə]	v. & n. 磨损，磨蚀，消耗，耗损
contaminate [kən'tæmɪneɪt]	n. 沾染，杂质，污染物质，污染剂
sealing ['siːlɪŋ]	n. 密封，封接
hostility [hɒ'stɪlɪtɪ]	n. 敌对，敌意，敌视
through-harden	v. 全部硬化，穿透淬火，淬透
HV=Vickers hardness	维氏硬度
notwithstanding [nɒtwɪð'stændɪŋ]	prep. 尽管，虽然；adv. 尽管，仍然
clearance ['klɪər(ə)ns]	n. 清除，空隙，清仓，间隙，清杆
interference [ɪntə'fɪər(ə)ns]	n. 干扰，冲突，干涉，妨碍，过盈
fret [fret]	v. 侵蚀，腐蚀，磨损，损坏
fretting [fretɪŋ]	n. 微振磨损，侵蚀
abut [ə'bʌt]	vt. 邻接，毗邻，紧靠，支撑；n. 尽头，支架
chamfer ['tʃæmfə]	n. & v. 在……开槽，倒角，圆角
swivel ['swɪv(ə)l]	n. 旋转轴承，转体

raceway ['reɪsweɪ]	n.	轴承座圈，滚道
lubrication [ˌluːbrɪ'keɪʃən]	n.	润滑剂，润滑材料
deterioration [dɪˌtɪərɪə'reɪʃn]	n.	恶化，退化，变质
tribology [traɪ'bɒlədʒɪ]	n.	摩擦学

⇨ Notes

[1] Rolling-contact bearings are generally made with steel which is through-hardened...
参考译文：滚动接触轴承一般用通过硬化的钢制成……。
through-hardended 意为"经过透硬淬火的，整体淬火的"。

[2] Inadequate interference can induce serious trouble from fretting corrosion.
参考译文：不充分的干扰会因为微动腐蚀而带来严重麻烦。
fretting corrosion 意为"微动腐蚀(即由腐蚀和两固体接触面有微小振幅的振动面引起的磨损之联合作用所产生的材料破坏)"。

[3] Thus the concept of "static load-carrying capacity" is understood to...
参考译文：因此，"静态承载能力"的概念被理解为……。
static load-carrying capacity 意为"静止承载能力"。

✗ Questions

1. What concerns should a machine designer with ball and roller bearings bear in mind?
2. What is the working principle of rolling contact bearings?
3. Under what circumstances are rolling contact bearings employed?
4. What are basic components of rolling contact bearings?
5. What advantages do rolling contact bearings have?

Reading Material: Bearings

A bearing is a connector that permits the connected members to either rotate or translate (move to and fro) relative to one another but prevents them from separating the direction in which loads are applied. In many cases one of the members is fixed, and the bearing acts as a support for the moving member.

The relative motion in bearings is always opposed by friction, and the work done in overcoming friction is lost power in all machines. Consequently, much thought and effort have been devoted to the development of bearing with minimum friction. In all bearings there are two surfaces (one belonging to each of the connected parts) that move relative to one another. To minimize friction, the co-acting surfaces may be partially or completely separated by a film of liquid or gas; these are known as sliding-contact bearings. The surfaces may be separated also by an assemblage of rolling elements such as balls and rollers; these are known as rolling-contact bearings.

Sliding bearings are the simplest to construct and, considering the multitude of pinpointed devices and structures in use, are probably the most commonly used.

The essential parts of a ball bearing—the inner and outer ring. The inner ring is mounted on a shaft and has a groove in which the balls ride. The outer rings are usually the stationary part of the bearing and also contain a groove to guide and support the balls. The separator prevents contact between the balls and thus reduces friction, wear, and noise from the regions where severe sliding conditions would occur. In a few applications where operating conditions are mild, the rings and separator can be omitted and loose balls interposed between the shaft and housing. This type of bearing is sometimes found in bicycles. There are many types of bearings because of variations in the design of rings and separators and in the number of balls. They can be divided into classes according to their function; those that support a radial load, those that support a thrust load, or those that support a combination of thrust and radial loads. The last type is termed "angular-contact bearings".

In deep-groove ball bearings, the races are approximately one fourth as deep as the ball diameter. A cross section of the ball bearings are designed to carry a radial load, they perform well under a combined radial and thrust load. For this reason, this is the most widely used type of ball bearing.

◇ New Words and Expressions

connector [kəˈnɛktɚ]	n. 连接器，接头
relative [ˈrelətɪv]	adj. 相对 (应、关) 的
permit [pəˈmɪt]	v. 许可，容 (允) 许
consequently [ˈkɒnsɪkw(ə)ntlɪ]	adv. 因此
devote [dɪˈvəʊt]	vt. 致力
prevent [priˈvɛnt]	v. 防止
reason [ˈriːz(ə)n]	n. 理由，原因
separate [ˈsep(ə)reɪt]	v. (使) 分离
apply [əˈplaɪ]	v. 适用 (合)，应用
fix [fɪks]	v. & n. (使) 固定，装配
omit [ə(ʊ)ˈmɪt]	vt. 省 (略) 去
loose [luːs]	adj. 松的
separator [ˈsepəreɪtə]	n. 分离 (分隔) 器
interpose	v. 放 (插、介) 入
groove [gruːv]	n. (凹、环、螺) 槽
guide [gaɪd]	vt. 指 (引、制) 导

region ['riːdʒ(ə)n]	n. 区 (域)；部位
mild [maɪld]	adj. 温和的，低碳的
(be) relative to	关于，相对于
angular-contact bearings	角接触球轴承
the vast majority	绝大多数
outer ring	外座图[环]
rolling-contact bearings	滚动轴承
sliding-contact bearings	滑动轴承
mild steel	低碳钢
co-acting	共同作用；相对运动
inner ring	内座图[环]

⊠ Questions

1. What is a bearing? What is its function?
2. Why have much thought and effort been devoted to the development of bearing with minimum friction?
3. What are the essential parts of a ball bearing?
4. What is the most widely used type of ball bearing?

Unit 15 Belts, Clutches, Brakes, and Chains

Belts, clutches, brakes, and chains are examples of machine elements that employ friction as a useful agent. A belt provides a convenient means for transferring power from one shaft to another. Belts are frequently necessary to reduce the higher rotative speeds of electric motors to the lower values required by mechanical equipment. Clutches are required when shafts must be frequently connected and disconnected. The function of the brake is to turn mechanical energy into heat. The design of frictional devices is subject to uncertainties in the value of the coefficient of friction that must necessarily be used. Chains provide a convenient and effective means for transferring power between parallel shafts.

The rayon and rubber V-belt is widely used for power transmission[1]. Such belts are made in two series: the standard V-belt and the high capacity V-belt. Other types of belts are available for power transmission purposes. The teeth of so-called timing belt will keep shafts completely synchronized[2].

When designing a V-belt drive it is a good plan to calculate the cost of two or three different layouts of belts and pulleys to determine which has the smallest overall cost. The catalogs of the various manufacturers of V-belts contain much practical information.

The V-belt is an important element in the field of power transmission. It is continually being improved by the various manufacturers and loading values are revised from time to time. The designer generally is guided by the current literature for the particular brand of belt he expects to use. The forgoing development should be considered a method that shows how the fatigue life of a belt is influenced by factors such as bending, centrifugal effects, and power transmitted.

A clutch is a device for quickly and easily connecting or disconnecting a rotatable shaft and a rotating coaxial shaft[3]. Clutches are usually placed between the input shaft to a machine and the output shaft from the driving motor, and provide a convenient means for starting and stopping the machine and permitting the driver motor or engine to be started in an unloaded state.

The rotor (rotating member) in an electric motor has rotational inertia, and a torque is required to bring it up to speed when the motor is started. If the motor shaft is rigidly connected to a load with a large rotational inertia, and the motor is started suddenly by closing a switch, the motor may not have sufficient torque capacity to bring the motor shaft up to speed before the windings in the motor are burned out by the excessive current demands [4]. A clutch between the motor and the load shafts will restrict the starting torque on the motor to that required to accelerate the rotor and parts of the clutch only. On some machine tools it is convenient to let the driving motor run continuously and to start and stop the machine by operating a clutch.

A brake is similar to a clutch except that one of the shafts is replaced by a fixed member. The basic function of a brake is to absorb energy (i.e., to convert kinetic and potential energy into friction heat) and to dissipate the resulting heat without developing destructively high temperatures[5]. Clutches also absorb energy and dissipate heat, but usually at a lower rate. Where brakes (or clutches) are used more or less continuously for extended periods of time, provision must be made for rapid transfer of heat to the surrounding atmosphere. For intermittent operation, the thermal capacity of the parts may permit much of the heat to be stored, and then dissipated over a longer period of time. Brake and clutch parts must be designed to avoid objectionable thermal stresses and thermal distortion.

The rate at which heat is generated on a unit area of friction interface is equal to the product of the normal pressure, coefficient of friction, and rubbing velocity[6]. Manufacturers of brakes and of brake lining materials have conducted tests and accumulated experience enabling them to arrive at empirical values of pV (normal pressure times rubbing velocity) and of power per unit area of friction surface that are appropriate for specific types of brake design, brake lining material, and service conditions.

Chain drives combine some of the more advantageous features of belt and gear drives. Chains provided almost any speed ratio for any practical shaft separation distance. Their chief advantage over gears is that chains can be used with arbitrary center distances. Compared with belts, chains offer the advantage of positive (no slip) drive and therefore greater power capacity[7]. An additional advantage is that not only two but also many shafts can be driven by a single chain at different speeds, yet all have synchronized motions. Primary applications are in conveyor systems, farm machinery, textile machinery, and motorcycles.

In its simplest form a chain drive consists of two sprockets of arbitrary size and a chain loop. Sprockets are wheels with external teeth shaped so that they can fit into the links of the drive or driven chain. The shape of the teeth varies with the number of teeth. In some recent automotive applications, tooth shape and/or size is modified to reduce noise generation.

Chains are available in a range of accuracies extending from precision to nonprecision. Nonprecision chains are low in cost and intended primarily for noncritical drives of less than 40-kW power ratings, precision chains, by contrast, are designed for high speeds and power capacity[8].

◇ New Words and Expressions

clutch [klʌtʃ]   n. 离合器
rotative ['rəʊtətɪv]   adj. 回转的，转动的
disconnect [dɪskə'nekt]   v. 拆开，使分离，断开
frictional ['frɪkʃənl]   adj. [力]摩擦的，由摩擦而生的
uncertainty [ʌn'sɜːt(ə)ntɪ]   n. 不确定，不可靠

friction ['frɪkʃ(ə)n]	n. 摩擦，[力] 摩擦力
rayon ['reɪɒn]	n. 人造丝，人造纤维
high-capacity V-belt	高负荷带
power transmission	动力传动装置，动力传递
foregoing ['fɔːgəʊɪŋ]	adj. 前述的，前面的，在前的
kinetic [kɪ'netɪk; kaɪ-]	adj. [力] 运动的
friction heat	摩擦热
dissipate ['dɪsɪpeɪt]	vt. 消散，使……消散
more or less	或多或少
extended period	持续时间，延长时间
surrounding atmosphere	周围空气，周围的大气
thermal capacity	热容量
lining ['laɪnɪŋ]	n. 衬里，内层，衬套
power capacity	功率，功率容量

⇨ Notes

[1] The rayon and rubber V-belt is widely used for power transmission.

参考译文：人造纤维和橡胶 V 带被广泛用来进行动力传送。

power transmission 意为"动力传送，动力传动装置"。

[2] The teeth of so-called timing belt will keep shafts completely synchronized.

参考译文：同步带上的齿可以使轴与轴之间实现完全同步。

timing belt 意为"同步带"。

[3] A clutch is a device for quickly and easily connecting or disconnecting a rotatable shaft and a rotating coaxial shaft.

参考译文：离合器是一个用来使从动轴与位于同一轴线上的主动轴进行快速和顺利的连接或脱开的装置。

rotating shaft 意为"转动的轴，主动轴"。

[4] If the motor shaft is rigidly connected to a load with a large rotational inertia, and the motor is started suddenly by closing a switch, the motor may not have sufficient torque capacity to bring the motor shaft up to speed before the windings in the motor are burned out by the excessive current demands.

参考译文：电机中的转子有转动惯量，当电机启动时需要一个转矩带动其运转。如果电动机的轴与具有很大转动惯量的负载刚性地连接在一起，当合上开关使电动机突然启动时，有可能在电动机没有来得及产生足够的扭矩，使电动机的轴达到应有的转速之前，电动机内的线圈就会因为过大的电流而被烧毁。

句中 rotational inertia 意为"转动惯量"。

[5] The basic function of a brake is to absorb energy (i.e., to convert kinetic and potential

energy into friction heat) and to dissipate the resulting heat without developing destructively high temperatures

参考译文：制动器的功能是吸收能量(即：将动能与势能转化为摩擦热) 并散热使不会达到破坏性的高温。

kinetic and potential energy 意为"动能与势能"。

[6] The rate at which heat is generated on a unit area of friction interface is equal to the product of the normal pressure, coefficient of friction, and rubbing velocity.

参考译文：在摩擦面之间单位面积上所产生热的速率等于正压力，摩擦系数和摩擦速度三者的乘积。

coefficient of friction 意为"摩擦系数"。

[7] Compared with belts, chains offer the advantage of positive (no slip) drive and therefore greater power capacity.

参考译文：与带传动相比，链传动的优点是强制(无滑动)传动，因此具有更大的传递动力的能力。

positive (no slip) drive 意为"强制(无滑动)传动"。

[8] Nonprecision chains are low in cost and intended primarily for noncritical drives of less than 40-kW power ratings, precision chains, by contrast, are designed for high speeds and power capacity.

参考译文：普通链的成本低，主要用于额定功率在 40 kW 以下的非关键性的传动中，与之相反，优质链被设计用于高速和传递大功率的场合。

power rating 意为"额定功率"。

Questions

1. Why are belts frequently necessary?
2. In what situation are clutches required?
3. What is the basic function of a brake?
4. Compared with belts, what advantage can chains offer?

Reading Material: Worm Gear Sets

Worm gear sets are widely used because of the many advantages obtained by action and load carrying capacity. A large speed reduction or a high increase of torque can be attained with the worm gear set. Compactness of design is easy to obtain with such a combination. Worm gear drives are quiet and vibration free [1].

A worm gear set consists of the worm, which is a helical gear. The shafts upon which the worm and the gear are mounted are usually at right angles but not in the same plane [2]. The usual practice is to have the worm drive the worm gear. With small ratios, it is impossible for the gear to drive the worm.

The worm can be made with either right-hand or left-hand threads. Also, like a screw, it can be made with single, double, triple, or quadruple threads. A single-threaded worm advances the worm gear a distance equal to pitch for each complete worm rotation. The distance advanced is called the lead. Thus, the pitch equals the lead for a single thread. A double-threaded worn has a lead which is twice the pitch.

The geometry of a worm is similar to that of power screw. Rotation of the worm simulates a linearly advancing involute rack. The geometry of worm gear (sometimes called a worm) is similar to that of a helical gear, except that the teeth are curved to envelop the worm. Sometimes the worn is modified to envelop the gear [3]. This gives a greater area of contact, but requires extremely precise mounting.

◈ New Words and Expressions

worm [wɜːm]	n. 蜗杆
pitch [pɪtʃ]	n. 节 (螺, 齿) 距
lead [liːd]	n. 导程, 螺距
geometry [dʒɪˈɒmɪtrɪ]	n. 几何 (学、图、形状)
torque [tɔːk]	n. 转 (动力) 矩, 扭矩
simulate [ˈsɪmjʊleɪt]	adj. 模拟的, 与……相仿的
attain [əˈteɪn]	v. 达到, 完成, 实现
similar [ˈsɪmɪlə]	adj. 相似的, 类似的
attain [əˈteɪn]	v. 达到, 完成, 实现
precise [prɪˈsaɪs]	adj. 精密的, 准确的
compactness [kəmˈpæktnɪs]	n. 致密, 紧凑
mount [maʊnt]	v. 安装
drive [draɪv]	n. 传动装置; v. 驱动, 传动
envelop [ɪnˈveləp; en-]	vt. 包 (装、围、络)
quiet [ˈkwaɪət]	adj. &. n. 安静, 平稳的
modify [ˈkwaɪət]	vt. (使) 变更改变
vibration	n. 振 (震、摆) 动
involute [ˈɪnvəl(j)uːt]	n. 渐开线
rack [ræk]	n. 齿条, 导轨
advance [ədˈvɑːns]	v. (使) 前进, 推进
worm gear	涡轮
tooth action	轮齿的 (相互) 作用; 啮合作用
load carrying capacity	传递载荷能力

(be)similar to...	与……相似，类似于
at right angle	相互垂直，互成直角
right-hand threads	右螺纹
left-hand threads	左螺纹

⇨ Notes

[1]　Worm gear drives are quiet and vibration free.

参考译文：蜗杆传动是安静和无振动的。

worm gear drive…and vibration free 中 free 是形容词，做后置定语，修饰名词"vibration"，译作"无振动"；同样的结构有 admission　free：免费入场。

[2]　The shafts upon which the worm and the gear are mounted are…same plane.

参考译文：安装蜗杆和涡轮的(两根)轴，不在同一平面中……。

upon which 引出主语的定语从句，该定语从句又是并列的定语从句，即 the worm 和 the gear 并列修饰这个 shafts，可译为"在轴之上"；而主句是 The　shafts…are　usually at right angle 译为"两根轴成直角"；but not in the same plane 是省略句；应为 but the shafts do not lie in the same plane 指两根轴"不在同一平面中"，而是处在"空间交错垂直"位置。

[3]　Sometimes the worm is modified to envelop the gears…

参考译文：有时把蜗杆齿纹制造成(弯弧状)包裹着涡轮。

the worm is modified 是被动语态，而 to envelop the gear 作主语 the worm 的补足语。

⊠ Questions

1. Why are worm gear sets widely used?
2. What is very similar to a screw?
3. Whose geometry is similar to that of power screw?
4. Whose geometry is similar to that of a helical gear?

PART 4 Machining Tools

Unit 16 Lathes

Lathes are machine tools designed primarily to do turning, facing, and boring. Very little turning is done on other types of machine tools, and none can do it with equal facility. Because lathes also can do drilling and reaming, their versatility permits several operations to be done with a single setup of the workpiece. Consequently, more lathes of various types are used in manufacturing than any other machine tool.

The essential components of a lathe are the bed, headstock assembly, tailstock assembly, carriage assembly, and the leadscrew and feed rod.

The bed is the backbone of a lathe. It is usually made of well-normalized or aged gray or nodular cast iron and provides a heavy, rigid frame on which all the other basic components are mounted. Two sets of parallel, longitudinal ways, inner or outer, are contained on the bed, usually on the upper side. Some makers use an inverted V-shape for all four ways, whereas others utilize one inverted V and one flat way in one or both sets. They are precision-machined to assure accuracy of alignment. On most modern lathes the ways are surface-hardened to resist wear and abrasion, but precaution should be taken in operating a lathe to assure that the ways are not damaged. Any inaccuracy in them usually means that the accuracy of the entire lathe is destroyed.

The headstock is mounted in a fixed position on the inner ways, usually at the left end of the bed; it provides a powered means of rotating the work at various speeds. Essentially, it consists of a hollow spindle, mounted in accurate bearings, and a set of transmission gears-similar to a truck transmission—through which the spindle can be rotated at a number of speeds. Most lathes provide form 8 to 18 speeds, usually in a geometric ratio, and on modern lathes all the speeds can be obtained merely by moving from two to four levers. An increasing trend is to provide a continuously variable speed range through electrical or mechanical drives.

Because the accuracy of a lathe is greatly dependent on the spindle, it is of heavy construction and mounted in heavy bearings, usually preloaded tapered roller or ball types. The spindle has a hole extending through its length, through which long bar stock can be fed. The size of this hole is an important dimension of a lathe because it determines the maximum size of bar stock that can be machined when the material must be fed through spindle.

The tailstock assembly consists, essentially, of three parts. A lower casting fits on the inner ways of the bed can slide longitudinally thereon, with a means for clamping the entire assembly in any desired location. An upper casting fits on the lower one and can be moved transversely upon it, on some type of keyed ways, to permit aligning the tailstock and headstock spindles. The third major component of the assembly is the tailstock quill. This is a hollow steel cylinder, usually about 51 to 76 mm in diameter, that can be moved several inches longitudinally in and out of the upper casting by means of a handwheel and screw.

The size of a lathe is designed by two dimensions. The first is known as the swing. This is the maximum diameter of work that can be rotated on a lathe. It is approximately twice the distance between the line connecting the lathe centers and the nearest point on the ways. The second size dimension is the maximum distance between centers. The swing thus indicates the maximum workpiece diameter that can be turned in the lathe, while the distance between centers indicates the maximum length of workpiece that can be mounted between centers.

Engine lathes are the type most frequently used in manufacturing. They are heavy-duty machine tools with all the components described previously and have power drive for all tool movements except on the compound rest. They commonly range in size from 305 to 610 mm (12 to 24 inches) swing and from 610 to 1219 mm (24 to 48 inches) center distances, but swings up to 1270 mm (50 inches) and center distances up to 3658 mm (12 feet) are not uncommon. Most have chip pans and a built-in coolant circulating system. Smaller engine lathes—with swings usually not over 330 mm (13 inches)—also are available in bench type, designed for the bed to be mounted on a bench or cabinet.

Although engine lathes are versatile and very useful, because of the time required for changing and setting tools and for making measurements on the workpiece, they are not suitable for quantity production. Often the actual chip-production time is less than 30% of the total cycle time. In addition, a skilled machinist is required for all the operations, and such persons are costly and often in short supply. However, much of the operation's time is consumed by simple, repetitious adjustments and in watching chips being made. Consequently, to reduce or eliminate the amount of skilled labor that is required, turret lathes, screw machines, and other types of semiautomatic and automatic lathes have been highly developed and are widely used in manufacturing.

◇ *New Words and Expressions*

lathe [leɪð]　　　　　　　　　n. & v. 车用，用车床加工，车削
turing ['tjuərɪŋ]　　　　　　　n. 旋转，车削，切削外圆
facing ['feɪsɪŋ]　　　　　　　n. 平面加工，端面加工
boring ['bɔːrɪŋ]　　　　　　　n. 镗孔，镗削加工
drilling ['drɪlɪŋ]　　　　　　　n. 钻孔
reaming ['riːmɪŋ]　　　　　　n. 铰孔
headstock ['hedstɒk]　　　　n. 头架，主轴箱
assembly [ə'semblɪ]　　　　n. 装配组合，组件，部件

tailstock ['teɪlstɒk]	n.	尾座，尾架，后顶尖座
carriage ['kærɪdʒ]	n.	(机床的) 拖板，机器的滑动部分
leadscrew [liːdskruː]	n.	丝杆
feed rod		光杠
way [wei]	n.	导轨
stock [stɒk]	n.	原料
lever ['liːvə]	n.	杠杆，手柄，把手
longitudinally [lɔŋdʒɪ'tjʊdɪnəli]	adv.	长度地，纵向地，轴向地
thereon [ðeər'ɒn]	adv.	在其中，在其上，关于那，紧接着
quill [kwɪl]	n.	活动套筒，衬套，钻轴，空心轴
swing [swɪŋ]	v. & n.	摇摆，摆动，最大回转半径
benchtype		台式
versatile ['vɜːsətaɪl]	adj.	多用途的，通用的，多方面适用的
turret ['tʌrit]	n.	转塔，六角刀架，六角车床

☒ Questions

1. What are the essential components of a lathe?
2. What parts does the tailstock assembly consist of?
3. Why are engine lathes not suitable for quantity production?
4. Where is the headstock mounted and what is its function?
5. What is the size of a lathe designed by?

Reading Material: Machine Tools

Machine tools are machines for cutting metals. The most important of machine tools used in industry are lathes, drilling machines and milling machines. Other kinds of metal working machines are not so widely used in machining metals as these three.

The lathe is commonly called the father of the entire machine tool family. For turning operations, the lathe uses a single-point-cutting tool, which removes metal as it travels past the revolving workpiece[1]. Turning operations are required to make many different cylindrical shapes, such as axes, gear blanks, pulleys, and threaded shafts. Boring operations are performed to enlarge, finish, and accurately locate holes.

Drilling is performed with a rotating tool called a drill. Most drilling in metal is done with a twist drill. The machine used for drilling is called a drill press. Operations, such as reaming and tapping, are also classified as drilling. Reaming consists of removing a small amount of metal from a hole already drilled.

Tapping is the process of cutting a thread inside a hole so that a cap screw or bolt may be threaded into it [2].

Milling removes metal with a revolving, multiple cutting edge tools called milling cutter. Milling cutters are made in many styles and sizes. Some have as few as two cutting edges and others have 30 or more. Milling can produce flat, angled surfaces, grooves, slots, gear teeth, and other profiles, depending on the shape of the cutters being used.

Shaping and planning produce flat surfaces with a single-point-cutting tool. In shaping, the cutting tool on a shaper reciprocates or moves back and forth while the work is fed automatically towards the tool[3]. The cutting tool is automatically fed into the work-piece a small amount on each stroke.

Grinding makes use of abrasive particles to do the cutting. Grinding operations may be classified as precision or non-precision, depending on the purpose. Precision grinding is concerned with grinding to close tolerances and very smooth finish. Non-precision grinding involves the removal of metal where accuracy is not important.

◇ *New Words and Expressions*

lathe [leɪð] n. 车床；v. 用车床加工
drill [drɪl] n. 钻床；v. 钻孔
mill [mɪl] n. 铣床，铣刀；v. 铣，铣削
ream [riːm] v. 铰孔
tap [tæp] v. 攻螺纹
thread [θred] n. 螺纹，线；v. 车螺纹
turn [tɜːn] v. 车削
cylindrical [səˈlɪndrɪkl] adj. 圆柱(体，形)的
blank [blæŋk] n. 毛坯，坯料
pulley [ˈpʊli] n. 皮带轮，滑轮
bore [bɔː(r)] n. 镗孔
enlarge [ɪnˈlɑːdʒ] v. 扩大
finish [ˈfɪnɪʃ] n. 光洁度；v. 精加工
shaper [ˈʃeɪpə] n. 牛头刨床
planer [ˈpleɪnə(r)] n. 龙门刨床
slot [slɒt] n. 缝，夹槽
reciprocate [rɪˈsɪprəkeɪt] v. 来回往复运动
stroke [strəʊk] n. 冲程，行程
abrasive [əˈbreɪsɪv] adj. 摩擦(损)的
particle [ˈpɑːtɪkl] n. 微粒
precision [prɪˈsɪʒn] n. 精度
consist of 包含，包括，由……组成

back and forth	来回，往复
be attached to	连接于，固定于
be concerned with	与……有关系
single point tool	单刃刀具

⇨ Notes

[1] For turning operations, the lathe uses a single-point-cutting tool, which removes metal as it travels past the revolving work-piece.

参考译文：为了进行车削，当工件旋转经过刀具时，车床用一把单刃刀具进行切削。

For turning operation 作状语，表目的，有"为了进行切削"之意。which 引导非限定性定语从句，修饰 a single-point-cutting tool。a single point cutting tool 切不可直译为"单点刀具"，而应该译为"单刃刀具"。

[2] Tapping is the process of cutting a thread inside a hole so that a cap screw or bolt may be threaded into it.

参考译文：攻螺纹就是在孔内壁上加工螺纹，以使螺钉或螺栓旋进孔内。

在英文翻译成中文时，不必把所有的英文一一对应翻译成中文。如原句中的 process 在译成中文时就不必译成"过程"，只要把句子的含义表达清楚就可以了。

[3] In shaping, the cutting tool on a shaper reciprocates or moves back and forth while the work is fed automatically towards the tool.

参考译文：用牛头刨床进行加工时，刀具在机床上往复运动，而工件朝向刀具自动进给。

用 or 连接的词或词组在地位上相等，有时表示的是同一个含义。本句中的 reciprocates 和 moves back and forth 表达同一含义，moves back and forth 是 reciprocates 的详细说明。

⌦ Questions

1. What are the most important machine tools used in industry?
2. What is a lathe? What is its function?
3. What is the function of shaping and planning?
4. Please explain what tapping is.
5. What is precision grinding and non-precision grinding respectively?

Unit 17 Drills and Drilling Machines

The twist dill is a very efficient tool. It is generally formed by forging and twisting grooves in a flat strip of steel or by milling a cylindrical piece of steel, high-speed steel being commonly used [1]. High-speed steel costs more but tools made of it withstand heat much better than those made of ordinary tool steel [2].

The twist drill may be divided into three principal parts: body, shank, and point. The flutes are the spiral grooves that are formed on the side of a drill, drills being made with two, or four flutes. Those having three or four flutes are used for following smaller drills or for enlarging holes already drilled, and are not suited for drilling into solid stock. (Figure 17-1).

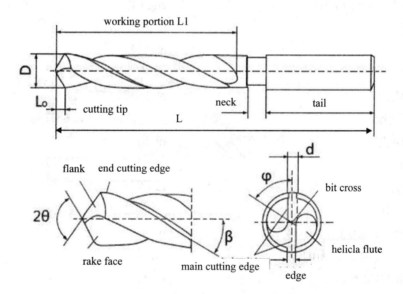

Figure17-1 The twist drill

Spiral flutes have four main advantages:
(1) They give the correct rake to the lip of a drill.
(2) They cause the chip to curl so tightly that it occupies the minimum amount of space [3].
(3) They form channels through which chips escape from the hole.
(4) They allow the lubricant to flow easily down to the cutting edge.

The margin is the narrow strip on the cutting edge of the flute. It is practically the full diameter of the drill and extends the entire length of the flute, its surface being a part of a cylinder [4]. The portion of the body next to the margin is of less diameter than the margin. This lessened diameter, called body clearance, reduces the friction between the drill and the walls of the hole, while the margin insures the hole being of accurate size.

The shank is the end of the drill which fits into the socket, spindle, or chuck of the drill press. The tang is usually found only on tapered shank tools.

The drilling machine is the second oldest machine tools, having been invented shortly after the lathe, and is one of the most common and useful machines [5]. The drilling machines may be classified into three general types: vertical spindle, multiple spindle, and radial spindle machines. The vertical spindle drilling machine comes in three types: heavy duty, plain, and sensitive [6].

Besides the drilling of holes, such operations may be performed on the drilling machine: drilling, tapping (internal threading), reaming (finishing the hole with a reamer), countersinking, counterboring, boring and spot-facing.

◇ *New Words and Expressions*

efficient [ɪˈfɪʃnt]	adj. 有效率的，能干的
principle [ˈprɪnsəpl]	adj. 主要的，首要的
body [ˈbɒdi]	n. 体，钻体
shank [ʃæŋk]	n. 钻柄
point [pɔɪnt]	n. 点，刀尖，钻尖
flute [fluːt]	n. 凹槽，螺旋槽
spiral [ˈspaɪrəl]	adj. 螺旋型的，螺纹的
rake [reɪk]	n. 前角
lip [lɪp]	n. 缘，刀刃，切削刃
chip [tʃɪp]	n. 碎屑，切屑
curl [kɜːl]	v. 使卷曲，使卷边
channel [ˈtʃænl]	n. 通道，沟槽
lubricant [ˈluːbrɪkənt]	n. 滑润剂
margin [ˈmɑːdʒɪn]	n. 刃带，边界，界限
lessen [ˈlesn]	v. 减少，减轻
socket [ˈsɒkɪt]	n. 钻套，插口
tang [tæŋ]	n. 柄舌
sensitive [ˈsensətɪv]	adj. 敏感的
tapping [ˈtæpɪŋ]	n. 攻丝，攻螺纹，穿孔
threading [ˈθredɪŋ]	n. 车螺纹
countersinking [ˈkaʊntəsɪŋkɪŋ]	n. 尖底锪钻
counterboring [ˈkaʊntɔːbərɪŋ]	n. 平底锪钻
boring [ˈbɔːrɪŋ]	n. 镗孔

spiral groove	螺旋槽
body clearance	钻体间隙
spot facing	锪端面

⇨ Notes

[1] It is generally formed by forging and twisting grooves in a flat strip of steel or by milling a cylindrical piece of steel, high-speed steel being commonly used.

参考译文：它通常由扁钢锻造后扭转出凹槽，或由圆柱形棒料铣削而成，一般由高速钢制作。

…，high-speed steel being commonly used 是被动形式的分词独立结构，它用来对句子的意思加以补充，可直接译成相当于汉语的一个句子。本句结构为"名词+being+过去分词"。在本课中类似的结构还有…，drills being made with two, three, or four flutes。

[2] High-speed steel costs more but tools made of it withstand heat much better than those made of ordinary tool steel.

参考译文：高速钢的成本虽然较高，但用它制作的刀具在耐热性方面要比用普通工具钢好得多。

[3] They cause the chip to curl so tightly that it occupies the minimum amount of space.

参考译文：(螺旋槽) 卷紧切屑，使其占有最小空间。

[4] It is practically the full diameter of the drill and extends the entire length of the flute, its surface being a part of a cylinder.

参考译文：(刃带部分) 实际上是钻头的最大直径处，它伸展到螺旋槽的全长，刃带的表面是圆柱体的一部分。

[5] The drilling machine is the second oldest machine tool, having been invented shortly after the lathe, …

参考译文：钻床是仅次于车床的最古老的机床，它的发明略晚于车床 (它在车床发明之后不久就被发明了)。

[6] The vertical spindle drilling machine comes in three types: heavy duty, plain, and sensitive.

参考译文：立式钻床有三种类型：重型钻床、普通钻床和高速手压台钻。

⊠ Questions

1. What parts may the twist drill be divided into?
2. What is the disadvantage of high-speed steel?
3. What advantages do spiral flutes have?
4. What is the drilling machine? When was it invented?
5. What types may the drilling machines be classified into?

Reading Material: Radial Drilling Machine

The radial drilling machine is designed for handling large work-pieces that cannot be easily moved. The drilling machine head is mounted on a heavy radial arm which may be from three to twelve or more feet long. This arm can be raised or lowered with power and can be turned in a complete circle around the column. The drilling head moves back and forth along this arm. On most radial drilling machines, movement of the arm, drill head, and spindle is controlled by power.

Spindle feeds and speeds are controlled by selector levers which engage the proper gears in the drill head. Depth of feed is also controlled directly in the drill head by a suitable machine. In addition to drills, other tools such as reamers and boring heads can also be used. Those tools add a great deal of versatility to this machine.

◇ New Words and Expressions

radial ['reɪdɪəl]	adj.	半径的
radially ['reɪdɪəlɪ]	adv.	沿径向
handle ['hændl]	n.	柄，把手；vt. 处理，操作
circle ['sɜːkl]	n.	圆周，圆；v. 环绕
control [kən'trəʊl]	n.	控制；vt. 控制
select [sɪ'lekt]	vt.	选择
selector [sɪ'lektə(r)]	n.	选择器
lever ['liːvə(r)]	n.	杆，杠杆
engage [ɪn'geɪdʒ]	vi.	接合，啮合
deal [diːl]	n.	量；vt. 分配
radial drilling machine		摇臂钻床
radial arm		摇臂
selector lever		选速手柄
a great deal of		大量的

⊠ Questions

1. What is the purpose of designing radial drilling machine?
2. Where is the drilling machine head mounted?
3. What are controlled by selector levers?
4. What other tools can also be used in addition to drills?

Unit 18 Milling, Shaper, Planer and Grinding Machines

The milling machine is a machine that removes metal from the work with a revolving milling cutter as the work is fed against it [1]. The milling cutter is mounted on an arbor where it is held in place by spacers or bushings. The arbor is fixed in the spindle with one end, while the other end of the arbor rotates in the bearing mounted on the arbor yoke.

The most important parts of the milling machine are: 1) starting levers; 2) spindle; 3) column; 4) knee; 5) elevating screw; 6) table; 7) index head; 8) speed levers; 9) feed levers; 10) table movement levers; 11) foot stock; 12) arbor yoke.

The spindle of the milling machine is driven by an electric motor through a train of gears mounted in the column.

The table of the plain milling machine may travel only at right angles to the spindle while the universal milling machine is provided with a table that may be swiveled on the transverse slide for milling gear teeth, threads, etc.

Sharper and Planer

The machine tools of this group are generally used for machining flat surfaces, which is usually performed by a cutter that peels the chip from the work. The main motion is reciprocating and the feed is normal (perpendicular) to the direction of the main motion.

The tool and the apron of a shaper are located on the ram. A chip is peeled off the work on the forward stroke. An adjustable table with "T"-slots holds the work, vise, and other fixtures for holding the work.

The shaper has a rocker arm which drives the ram, and a mechanism for regulating the length of the stroke. The ram supports the tool head. The head carries the downfeed mechanism and will swivel from side to side to permit the cutting of angles. This is generally a hand feed, but some shapers are equipped with a power downfeed in addition to the regular hand down feed.

The table of the shaper is of box form with "T"-slots on the top and sides for clamping the work. The cross rail is bolted directly to the frame or column of the shaper with bolts.

The automatic feed or power feed is obtained by a pawl which engages in a wheel or ratchet.

◇ **New Words and Expressions**

milling machine　　　铣床
milling cutter　　　　铣刀

⇨ **Notes**

[1]　The milling machine is a machine that removes metal from the work with a revolving milling cutter as the work is fed against it.

参考译文：铣床是一种铣刀旋转、工件相对作进给运动的金属切削机床。

that 引导的是限定性定语从句。

✕ **Questions**

1. What are the features of a milling machine?
2. What are the most important parts of the milling machine?
3. How can an electric motor drive a milling machine?

Reading Material: Milling Machines and Grinding Machines

Milling machines may be classified in three groups: the bed type, the column and knee type, and special milling machines [1].

The bed type milling machine is used for production manufacturing. The general characteristics of this type of machine are that the height of the table is fixed and the adjustments for height are made with the spindles [2]. This height adjustment is made by moving the spindle head up or down.

In construction, these machines are very rigid, permitting greater metal removal. Since adjustments are rather time consuming, it takes longer to set them up than it takes to set up the knee and column type of machine.

The planer type of milling machine is bed type machine which is usually very large. The table moves under an arch very much as the table of a planer. The cutter heads are mounted above and at the sides of the table [3].

The column and knee type of milling machine is the most commonly used because of its flexibility. Because of the ease with which it can be set up and its versatility, it is more adaptable for quick single-piece setups. If more complicated setups are desirable, this is also possible.

In general, these are two types of column and knee milling machines: the horizontal and vertical milling machines [4]. The difference between the horizontal and vertical milling machine is position of the spindle in relation to the machine table. The spindle is mounted vertically on the vertical miller and horizontally on the horizontal miller.

There are also several special-purpose machines such as thread millers, tracer-controlled (profilers) milling machines, rotary millers, and many others each designed to do a very special job.

The horizontal boring mill is usually a large machine designed with a retractable spindle mounted in the headstock of the machine. The headstock can be moved in a vertical direction. A bushing mounted in a block in the tailstock is caused to move up and down, always in alignment with the headstock spindle. Thus a long boring bar, capable of mounting cutters, can be supported at both ends. The boring bar receives its power from the spindle. It should be noted that milling cutters, drills, and so on, can be mounted directly in the spindle without using the boring bar or the tailstock.

Special milling machines for continuous milling are also manufactured. The rotary miller, which uses a vertical spindle and a power-actuated rotary table, provides an opportunity for continuous milling. Many pieces of work may be mounted in a circle. Since the rotation of the table is slow, the operator has the chance to remove the finished workpiece and load an unfinished workpiece.

There are many types of grinding machines. The two most widely used are the cylindrical and surface grinders[5]. Other grinding operations include those using vertical spindle, internal and centerless grinders, and jigs.

The cylindrical grinder is equipped with a headstock and tailstock between which the work is mounted and rotated as a grinding wheel in contact with the work removes metal from its circumference (Figure 18-1). The operation is somewhat similar to that of a lathe. The grinding wheel replaces the tool bit, and the thousands of little abrasive particles in it may be thought of as little tool bits. They in fact produce little chips during the cutting operation.

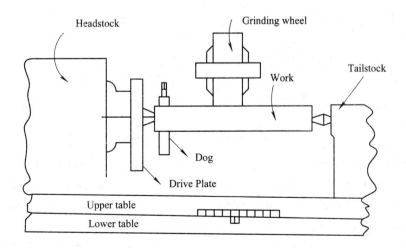

Figure 18-1 Sketch of cylindrical grinder

The headstock center may or may not revolve with the work. The tailstock center is always dead[6]. Operating a cylindrical grinder with both centers dead eliminates any possible eccentricity which may result from the live center runout. Precision grinding is done between two dead centers.

Feeding the grinding wheel into the work may be done automatically or by hand in increments as low as 0.0001 in. per pass. The range of infeed of the wheel into the work is from about 0.002 in. per pass for roughing to about 0.0005 in. for finishing. Hydraulically controlled infeed grinders are also available with mechanisms for automatic table retraction when the grinding operation is complete.

The surface grinder is used for grinding flat surfaces (Figure 18-2(a)). The table mounts a magnetic chuck used for holding the work during the grinding operation. The table oscillates under the wheel, as shown in Figure 18-2(b). With each pass the table feeds transversely. This feed may be accomplished automatically or by hand. The automatic control may be with hydraulic power or a mechanical indexing mechanism. As the table moves longitudinally the direction is reversed with trip dogs [7].

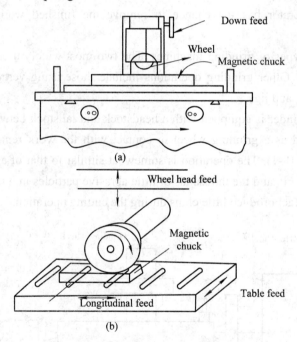

Figure 18-2 Sketch of a surface grinder

A unique grinder is the centerless grinder of which there are several types: the cylindrical grinders for grinding circular external surfaces, internal centerless grinders for grinding internal circular surfaces, and centerless thread grinders. These grinders may be used to grind cylinders, tapers, spheres, threads, and so on.

In all case at least three points of contact with the work are employed. The work is supported by a blade, or a roller. A regulating wheel which has a higher coefficient of friction than the grinding wheel is the second support point[8]. The grinding wheel is the third support point. This is shown in Figure 18-3.

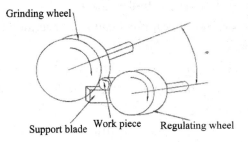

Figure 18-3 Sketch of a centerless grinder

◆ *New Words and Expressions*

longitudinal [ˌlɒŋgɪ'tjuːdɪnl] adj. 经度的，纵向的
milling ['mɪlɪŋ] n. 磨，制粉，[机] 轧齿边
grinding ['graɪndɪŋ] adj. 磨的，摩擦的，碾的
knee [niː] n. 合角铁，(铣床等的) 升降台
flexibility [ˌfleksə'bɪlətɪ] n. 弹性，适应性，机动性，挠性
versatility [ˌvɜːsə'tɪlətɪ] n. 多功能性
horizontal [ˌhɒrɪ'zɒntl] adj. 地平线的，水平的
retractable [rɪ'træktəbl] adj. 可缩进的，可缩回的
bushing ['bʊʃɪŋ] n. [机] 轴衬，[电工] 套管
alignment [ə'laɪnmənt] n. 队列，结盟
poweractuated [paʊə'ræktʃʊeɪtɪd] adj. 电力驱动的，机械传动的，机动的
jig [dʒɪg] n. 快步舞；带锤子的钓钩，夹具
abrasive particle [ə'breɪsɪv'pɑːtɪkl] 磨粒，磨料微粒
eccentricity [ˌeksen'trɪsətɪ] n. 偏心；古怪；[数] 离心率
increment ['ɪŋkrəmənt] n. 增加，增量
hydraulic [haɪ'drɔːlɪk] adj. 水力的，水压的

⇨ *Notes*

[1] Milling machines may be classified in three groups: the bed type, the column and knee type, and special milling machines.

参考译文：铣床可以被分为三种类型：床身式铣床、升降台式铣床和专用铣床。

[2] The general characteristics of this type of machine are that the height of the table is fixed and the adjustments for height are made with the spindles.

参考译文：这种机床的一般特性是工作台的高度是固定的，高度的调整是通过主轴进行的。

[3] The cutter heads are mounted above and at the sides of the table.

参考译文：刀盘安装在工作台的上方和侧面。

[4] In general, these are two types of column and knee milling machines: the horizontal and vertical milling machines.

参考译文：通常，有两种升降台式铣床：卧式铣床和立式铣床。

[5] There are many types of grinding machines. The two most widely used are the cylindrical and surface grinders.

参考译文：有许多种类的磨床，两种最常用的磨床是外圆磨床和平面磨床。

[6] The tailstock center is always dead.

参考译文：尾架总是采用死顶尖。

[7] As the table moves longitudinally the direction is reversed with trip dogs.

参考译文：工作台在纵向移动中碰到行程挡块时，就会改变运动方向。

[8] The work is supported by a blade, or a roller. A regulating wheel which has a higher coefficient of friction than the grinding wheel is the second support point.

参考译文：工件被支承在托板或辊子上。一个具有比砂轮摩擦系数更高的导轮是第二个支撑点。

✗ Questions

1. In what groups can milling machines be classified?
2. What are the general characteristics of bedtype milling machine?
3. Why is the column and knee type of milling machine used commonly?
4. What types of grinding machines are the most widely used?

Unit 19 Machine Tool Tests, Accuracy Checking and Maintenance

Machine Tool Tests and Accuracy checking

After manufacture or repairs, each machine tool should meet the requirements of specifications. According to the approved general specifications, acceptance tests of machine tools should include:

(a) idle-run tests, mechanisms operation checks, certificate data checks;
(b) load tests and productive output tests (for special machine tools);
(c) checks of the geometrical accuracy, surface roughness, and accuracy of the workpiece being machined;
(d) rigidity tests of machine tools;
(e) tests for vibration-proof properties of machine tools cutting;

These tests of machine tools should be conducted in the above sequence. The accuracy of the workpiece being machined and its surface roughness may be checked during the load tests of the machine tool and before the geometrical accuracy of the latter is checked.

Accuracy checks are considered below. These include the checking of the machine tool geometrical accuracy, the accuracy of the workpiece machined and its surface roughness. Machine tool geometrical accuracy tests include: the checking of guide ways for straightness; work tables for flatness; columns, uprights, and base plates for deviation from the vertical and horizontal planes; spindles for correct location and accuracy of rotation; relative position of axes and surfaces for parallelism and squareness; lead screws and indexing devices for specific errors; etc. These checks are conducted in accordance with the GOSTs for a given type of machine tool.

Geometrical accuracy tests alone are inadequate to judge the machine-tool performance because they do not (or inadequately) reveal variations in rigidity of machine-tool components, the quality of their manufacture and assembly, and especially, the influence of the machine-fixture-cutting tool-workpiece system rigidity on the accuracy of machining. That is why the corresponding State Standards stipulate compulsory accuracy tests of machine tools by machining work samples including a check of their surface roughness. These tests should be carried out after the preliminary idle running of the machine tool or its load tests, with essential parts of the machine having a stabilized working temperature [1]. The king of work sample, its material, and the character of machining for various types of machine tools are given in corresponding standards.

Machine Tool Maintenance

The maintenance of machine tools in the USSR is accomplished in accordance with a planned maintenance system(PMS).This system provides for complex measures for servicing, inspection, and repairs of equipment, which prevent the wear of equipment and help to keep it in good order[2]. The PMS system can be affected by means of the following methods:

(1) Post-inspections repairs. This method involves the planning of periodic inspections rather than repairs. The time interval between successive inspections is determined according to the minimum service life of rapidly wearing components. If an inspection confirms that there is no need for repairs and that the machine can operate without these until the next inspection, the repairs are postponed. This method prevents any sudden breakdown of equipment.

(2) The method of periodic repairs consists in repairs being done after a given running time.

(3) The method of standard (compulsory) repairs involves compulsory repairs of equipment at planned intervals, which are standard for each piece of equipment.

The PMS system includes:

(1) Routine servicing, which provides for normal everyday running of machines, minor repairs and, whenever necessary, adjustment of separate units or members of the machine tool.

(2) Periodic inspections, which are conducted according to schedule and involve visual inspection cleaning, and accuracy checks.

(3) Inspections as such consist in exterior checks accompanied by partial disassembly. All the mechanisms are checked in operation and regulated; fasteners are repaired or replaced; the state and wear of the machine tool as a whole and its individual units are assessed. The inspection results are recorded in a report on the mechanical condition of the machine. The date for the next repairs is defined in accordance with this report. The accuracy check of the machine tool is conducted according to approved standards.

(4) Scheduled repairs are divided into minor (or routine), medium, and general repairs. In routine repairs, separate components or units of the machine are repaired or replaced without thorough disassembly of the machine. Medium repairs include all the elements of routine repairs with the additional restoration of the relative position of the principal units and with partial repairs to basic components.

General overhaul involves the complete replacement or repairs of all the basic components, full restoration of the relative position of the principal units and the required accuracy of the machine tool.

◇ *New Words and Expressions*

straightness ['streitnis]   n. 平直度
flatness ['flætnɪs]   n. 平面度
upright ['ʌprait]   n. 立柱

Unit 19　Machine Tool Tests, Accuracy Checking and Maintenance

baseplate ['beɪspleɪt]	n.	底板
deviation [ˌdiːviˈeɪʃən]	n.	偏差
parallelism ['pærəlelizəm]	n.	平行度
squareness ['skweənɪs]	n.	垂直度
servicing ['səːvisiŋ]	n.	维修
inspection [ɪnˈspekʃən]	n.	检查
repair [riˈpɛə]	n.	检修
maintenance ['meintinəns]	n.	保养
restoration [ˌrestəˈreɪʃən]	n.	修复
acceptance tests		验收试验
certificate data		检验证数据
geometrical accuracy		几何精度
surface roughness		表面粗糙度
vibration-proof		防振
rapidly wearing components		易损件
post-inspections repairs		后检查修理
periodic repairs		定期检修
routine servicing		日常检修
general overhaul		大修

⇨ Notes

[1]　These tests should be carried out after the preliminary idle running of the machine tool or its load tests, with essential parts of the machine having a stabilized working temperature.

参考译文：这些试验应在机床最初空转或负载试验后，机床主要部件处于稳态工作温度时进行。

介词 with 后面有时跟一个名(代)词和另一个成分构成的复合结构。这种介词短语可以做状语，表示背景情况或行为方式，原因或条件。

[2]　This system provides for complex measures for servicing, inspection, and repairs of equipment, which prevent the wear of equipment and help to keep it in good order.

参考译文：这些系统可对设备的维修、检查及修理进行综合测量，从而可防止设备的磨损，有助于设备的正常运行。

⊠ Questions

1. How can we check the accuracy of a machine tool?
2. How can we repair a machine tool?
3. Why are geometrical accuracy tests alone inadequate to judge the machine-tool performance?

Reading Material: Physical Basis of the Cutting Process

Preliminary Information

The removal of the undeformed chip, its transformation into actual chips, and the production of the machined surface, is a complex physical(and, partly, chemical)process. During elastic and plastic deformation several complex processes such as heat generation and propagation, internal and external friction, wear, hardening and tempering etc, are simultaneously taking place.

Research in cutting takes place in order to understand the nature and course of all those phenomena which tend to a rational application and control of the cutting process [1]. The study of the nature and course of the complex cutting process is usually achieved by means of observation, research and analysis of phenomena and processes using simplified models of the actual cutting process.

Orthogonal Cutting as a Case of Plane Plastic Deformation

A form of cutting which can be considered as a case of plane deformation is called orthogonal cutting. It takes place when cutting is carried out by means of a single rectilinear edge which is longer than the width of cut and the width of cut is sufficiently great and considerably greater than the undeformed chip thickness.

The mechanics of the orthogonal cutting process is considered using the following assumptions:

—the properties of the machined material are isotropic;
—the temperature does not change in the whole cutting area;
—the machined material hardens(becomes stronger)as a result of permanent deformation;
—the condition of plasticity is the appearance of maximum shear stresses;
—the portion of the tool point, which adjoins the flank, does not take part in the process of plastic deformation, and the radius of the cutting edge is equal to zero.

The orthogonal cutting process can be considered analogous to the process of pressing a punch into a piece of material. In such a case deformation lines appear in the pressed material. If it is assumed that the compression is not accompanied by friction, then the orthogonal network of deformation lines will form an angle of 45° with the flat surface of the punch [2]. In the area OBC the deformation network will consist of a number of radii starting from point 0 and of concentric arcs.

A simplified diagram of the area and lines of plastic deformation in the orthogonal cutting process is limited by:

—line OP called the initial boundary of plastic deformation,
—line PK called the external boundary of plastic deformation,

—line KO called the final boundary of plastic deformation

The area within these boundaries is the area where the undeformed chip becomes the actual chip, or the area of chip formation.

The position of the boundary curves, and particularly that of the initial boundary of plastic deformation, was established by theoretical analysis and confirmed experimentally by a number of research workers. It is important to note that the initial curve OP, in its lower part, is situated below the line of cutting. This indicates that the area of plastic deformation reaches the material not transformed into the chip.

It follows from the system of deformation lines that shear does not take place on a single plane[3]. Compressive stresses change within the chip formation area from the value corresponding to the initial yield stress of the material to the value corresponding to the yield stress of the hardened (strengthened) chip material. Plastic deformation also increases correspondingly. The boundaries of the chip formation area and the chip deforming process itself are not stable and uniform. A study of the chip formation process indicates a periodic increase in stress and deformation, accompanied by slip[4]. As a result, on the external surface of the chip there appear creases and shoulder-like irregularities. This phenomenon is more distinct when the deformed material is less plastic. The periodic increase in stress, deformation and slip produces the elementary chip structure observed by the early researchers on the cutting process.

Characteristics of Non-orthogonal Cutting

Orthogonal cutting seldom occurs in industrial practice, and, for production purposes, non-orthogonal cutting is commonly applied.

A non-orthogonal machining process occurs when cutting:

—is carried out simultaneously by two or more rectilinear cutting edges or else by a curvilinear edge of the tool point;

—cannot be considered as a plane deformation process.

Whilst there are differences caused in the non-orthogonal process by simultaneous work of two or several cutting edges, all physical phenomena occur in a qualitatively similar way [5]. Therefore, for the purpose of explaining and understanding those phenomena and the laws governing them, the simpler model of orthogonal cutting is used.

When evaluating the non-orthogonal cutting process in a quantitative manner the same basic definitions should be used, but at the same time the fact that all active sections of the cutting edge may work simultaneously should be taken into consideration.

Quantitative differences between non-orthogonal cutting and those concerning orthogonal cutting increase with the ratio of the length of trail edge to that of leading cutting edge.

Therefore, if this ratio is smaller than 0.1 then, from the point of view of the mechanics of the cutting process, the reactions valid for orthogonal cutting [6] approximately apply to the non-orthogonal case.

Chip Texture and Built-up Edge

When cutting is dry, the chip moving along the rake face must overcome the frictional force. The latter takes place under specific conditions:

—the normal pressure may reach several thousand kilograms force per square cm;

—mating surfaces, and particularly the cutting surfaces, are clean in the physical and chemical sense,

—contact surface are in a non-uniform temperature field and maximum temperature at the contact area may reach the melting temperature of the material being machined.

The frictional force of the chip against the rake face depends on the area of the contact surface and on the distribution of normal pressure and coefficient of friction over that surface.

Experiments conducted by many research workers indicate that the coefficient of friction may considerably exceed 0.5. Principal factors influencing the value of the coefficient of friction are(besides the properties of materials rubbing against each other): the temperature of the contact surface, the medium in which cutting takes place, the normal pressure and tool point rake angle.

As a result of friction on the contact surface between chip and tool point, the movement of chip layers nearest to the contact surface is delayed.

Under certain conditions prevailing on the surface of contact between chip and the rake face, friction may increase so much(on a smaller or larger part of the face)that the movement of further layers of the chip being formed is delayed[7]. This results from the external friction stress exceeding the yield stress of the nearest layers of the moving chip[8]. A characteristic texture of the bottom layers of the chip then appears which can be detected by metallographic methods.

Besides chip texture which is always more or less evident, under certain conditions of cutting ductile metals the appearance of the so-called built-up edge (BUE) can be observed. This is a wedge-like extension of the edge, consisting of the machined material.

The built-up edge partially takes over the work of the tool point, at the same time not only changing the process of chip formation, but also(which is more important)changing the dimensions and character of the machined surface. Therefore knowledge of the law governing this building-up process is of great importance.

Plastic deformation and external friction of the chip against the tool point generate a certain amount of heat. This results in the formation of a high temperature field which also comprises the chip formation zone.

The high temperature field is characterized by a high gradient and therefore considerably modifies the physical and mechanical properties of the chip formation area.

According to observations and experiments carried out, conditions may be created in which, notwithstanding the influence of the high temperature field, the state of plasticity cannot be maintained in the area of maximum compression. As a result, the harder, wedge-shaped part of the machined material, adhering to the tool point because of high friction, cuts like an extended edge into the plastic workpiece material. This phenomenon occurs in a continuous way.

Thus the built-up edge gradually become thicker. The condition necessary for the BUE to adhere to the tool point is a sufficiently high coefficient of external friction (in principle, higher than 0.5) and such a system of forces acting upon the tool point that does not exceed the strength or the frictional grip of the BUE on the tool point.

As the BUE increases on the tool point, so does the friction of the chip material flowing around it, and so do the forces and moments acting upon the BUE.

If at first the frictional resistance around the built-up edge (in the direction of chip flow) becomes equal to the friction of the BUE against the rake-face, then the entire BUE flows along with the chip. If at first the strength of the BUE is exceeded, which may cause its splitting, breaking or shearing off, then isolated parts of the BUE may be retained by the chip and pressed into the machined material.

In both cases the formation or growth of the BUE is recommenced after it has broken away.

The result of the BUE is as if a new tool point, with a different rake angle, as created. This angle is called the BUE rake angle γ_u or, shortly, the BUE angle. It is greater than the rake angle.

The dimensions of BUE are characterized by:

—BUE height h_u,

—BUE overhang υ which to a large extent determines the irregularities of the machined surface.

The boundaries of the BUE may be determined from photomicrographs or by hardness testing. At the boundary of the BUE there is a sudden increase in hardness, i.e. the BUE hardness is higher than that of the chip which, in turn, is harder than the machined material.

Built-up edge has a great and, taking all circumstances into consideration negative influence upon the course and results of the cutting process on chip formation and flow. BUE has an influence similar to that of an increase in rake angle. The larger the built-up edge, the smaller is the chip area factor and the smaller becomes the cutting force. BUE also reduces frictional wear of the tool point, but on the other hand, by periodic appearance and disappearance, accompanied by changes of cutting force value, may become a cause of vibrations. This in turn may cause an increase of fatigue wear of the tool point. First of all, however, BUE negatively influences the quality of the machined surface and this is the main reason why cutting under conditions permitting the formation of BUE is considered to be abnormal and disadvantageous.

Knowledge of the laws governing the appearance and disappearance of BUE should be used to the purpose of avoiding the machining conditions favouring its appearance.

◇ New Words and Expressions

medium ['mi:diəm]	n. 介质
adjoin [ə'dʒɔɪn]	v. 毗连
along with	与……一道
boundary ['baundəri]	n. 边界，界面

take over	接任
built-up edge	积削瘤
follow from	根据……得出
or else	否则，要么，要不就
in particular	特别是
yield stress	屈服应力
for the purpose of	为了……
texture ['tekstʃə(r)]	（金相）组织
take into consideration	考虑到
orthogonal cutting	直角切削
carry out	执（进）行

➪ Notes

[1] Research...of the cutting process.

参考译文：对切削加工进行研究，目的在于了解上述现象的发生过程及本质，以便能合理运用和控制它们。

which 引导的定语从句说明 the nature and course of all those phenomena。

[2] If it is assumed that...of the punch.

参考译文：若假设在挤压的同时没有摩擦产生，则互为直角的应力线簇和冲头平面成 45° 角。

这里 that 引导的是主语从句。

[3] It follows from... a single plane.

参考译文：由图示应力线簇推断，剪切变形不止在一个平面里进行。

It follows (from...)that...可译为"根据……推断……"；that 引导的是主语从句。

[4] A study ... by slip.

参考译文：关于切削变形过程的研究表明，变形和应力是周期性增加的，并有滑移伴随产生。

A study of the chip formation process 是名词性词组，代替一个句子的意思，以便使文字简练，其结构形式是 n.+ of-phrase +other modifiers，其中 n. 常为表示动作意义的名词，of-phrase 表示行为的发出者或是行为的对象。

[5] Whilst there are...similar way.

参考译文：虽然在斜角切削过程中由于两刃或多刃同时参加切削会引起差异，但从本质上讲，其基本物理现象与直角切削是相似的。

[6] the reactions valid for orthogonal cutting.

参考译文：用于正交切削的反应。

valid for orthogonal cutting 是形容词短语做后置定语，说明 reactions。

[7] Under certain conditions prevailing on the surface of contact between chip and the rake

face, friction may increase so much(on a smaller or larger part of the face)that the movement of further layers of the chip being formed is delayed.

参考译文：在一定条件下，切削和前刀面之间的摩擦在接触面的某个区域可能会大到使刚变形的切削不能继续流动。

句中 prevailing on...rake face 为分词短语，作定语。

[8] This results from the external friction stress exceeding the yield stress of the nearest layers of the moving chip.

参考译文：这是由外部摩擦应力超过了最近的移动芯片层的屈服应力所导致的。

其中 the external frictional stress exceeding the...chip 是动名词复合结构作介词 from 的宾语，the external frictional stress 是 exceeding 的逻辑主语。

⊠ Questions

1. What is the purpose of research in cutting?
2. What is orthogonal cutting?
3. What is the function or purpose of orthogonal cutting?
4. What are characteristics of non-orthogonal cutting?
5. When does a non-orthogonal machining process occur?

Unit 20   Nontraditional Manufacturing Processes

The human race has distinguished itself from all other forms of life by using tools and intelligence to create items that serve to make life easier and more enjoyable. Through the centuries, both the tools and the energy sources to power these tools have evolved to meet the increasing sophistication and complexity of mankind's ideas.

In their earliest forms, tools primarily consisted of stone instruments. Considering the relative simplicity of the items being made and the materials being shaped, stone was adequate. When iron tools were invented, durable metals and more sophisticated articles could be produced. The twentieth century has seen the creation of products made from the most durable and, consequently, the most difficult-to-machine materials in history. In an effort to meet the manufacturing challenges created by these materials, tools have now evolved to include materials such as alloy steel, carbide, diamond, and ceramics.

A similar evolution has taken place with the methods used to power our tools. Initially, tools were powered by muscles; either human or animal. However as the powers of water, wind, steam, and electricity were harnessed, mankind was able to further extended manufacturing capabilities with new machines, greater accuracy, and faster machining rates.

Every time new tools, tool materials, and power sources are utilized, the efficiency and capabilities of manufacturers are greatly enhanced. However as old problems are solved, new problems and challenges arise so that the manufacturers of today are faced with tough questions such as the following: How do you drill a 2 mm diameter hole 670 mm deep without experiencing taper or runout? Is there a way to efficiently deburr passageways inside complex castings and guarantee 100% that no burrs were missed? Is there a welding process that can eliminate the thermal damage now occurring to my product?

Since the 1940s, a revolution in manufacturing has been taking place that once again allows manufacturers to meet the demands imposed by increasingly sophisticated designs and durable, but in many cases nearly unmachinable, materials. This manufacturing revolution is now, as it has been in the past, centered on the use of new tools and new forms of energy. The result has been the introduction of new manufacturing processes used for material removal, forming, and joining, known today as nontraditional manufacturing processes.

The conventional manufacturing processes in use today for material removal primarily rely on electric motors and hard tool materials to perform tasks such as sawing, drilling, and broaching. Conventional forming operations are performed with the energy from electric motors, hydraulics, and gravity. Likewise, material joining is conventionally accomplished with thermal energy sources such as burning gases and electric arcs [1].

In contrast, nontraditional manufacturing processes harness energy sources considered unconventional by yesterday's standards. Material removal can now be accomplished with electrochemical reactions, high-temperature plasmas, and high-velocity jets of liquids and abrasives [2]. Materials that in the past have been extremely difficult to form, are now formed with magnetic fields, explosives, and the shock waves from powerful electric sparks. Material-joining capabilities have been expanded with the use of high-frequency sound waves and beams of electrons.

In the past 50 years, over 20 different nontraditional manufacturing processes have been invented and successfully implemented into production. The reason there are such a large number of nontraditional processes is the same reason there are such a large number of conventional processes; each process has its own characteristic attributes and limitations, hence no one process is best for all manufacturing situations.

For example, nontraditional process are sometimes applied to increase productivity either by reducing the number of overall manufacturing operations required to produce a product or by performing operations faster than the previously used method.

In other cases, nontraditional processes are used to reduce the number of rejects [3] experienced by the old manufacturing method by increasing repeatability, reducing in-process breakage of fragile workpieces, or by minimizing detrimental effects on workpiece properties.

Because of the aforementioned attributes, nontraditional manufacturing processes have experienced steady growth since their introduction. An increasing growth rate for these processes in the future is assured for the following reasons:

1. Currently, nontraditional processes possess virtually unlimited capabilities when compared with conventional processes, except for volumetric material removal rates. Great advances have been made in the past few years in increasing the removal rates of some of these processes, and there is no reason to believe that this trend will not continue into the future.

2. Approximately one half of the nontraditional manufacturing processes are available with computer control of the process parameters. The use of computers lends simplicity to processes that people may be unfamiliar with, and thereby accelerates acceptance. Additionally, computer control assures reliability and repeatability[4], which also accelerates acceptance and implementation.

3. Most nontraditional processes are capable of being adaptively controlled through the use of vision systems, laser gages, and other in-process inspection techniques. If, for example, the inprocess inspection system determines that the size of holes being produced in a product are becoming smaller, the size can be modified without changing hard tools, such as drills.

4. The implementation of nontraditional manufacturing processes will continue to increase as manufacturing engineers, product designers, and metallurgical engineers become increasingly aware of the unique capabilities and benefits that nontraditional manufacturing processes provide.

◇ *New Words and Expressions*

sophistication [sə,fɪstɪ'keɪʃn]	n. 复杂化，完善，采用先进技术
durable ['djʊərəb(ə)l]	adj. 耐用的，持久的
alloy ['ælɒɪ]	n. 合金；v. 熔合
carbide ['kɑːbaɪd]	n. 碳化物，硬质合金
taper ['teɪpə]	n. 圆锥，锥度，逐渐缩减
runout ['rʌnaut]	n. 避开，偏摆，径向摆动，径向跳动
deburr [diːˈbəː]	vt. 去毛刺
passageway ['pæsɪdʒweɪ]	n. 通道，走廊
casting ['kɑːstɪŋ]	n. 铸造，铸件
burr [bɜː]	n. 毛刺
thermal ['θɜːm(ə)l]	adj. 热的，热力的
sawing ['sɔːiŋ]	n. 锯切，锯开
unconventional [ʌnkən'venʃ(ə)n(ə)l]	adj. 非常规的，非传统的，不依惯例的
electrochemical [i,lektrəu'kemikəl,-'kemik]	adj. 电化学的，电气化学的
reject [rɪ'dʒekt]	vt. 拒绝，排斥，丢弃；n. 不合格品，废品
breakage ['breɪkɪdʒ]	n. 破坏，破损，裂口，破损量
fragile ['frædʒaɪl]	adj. 脆的，易碎的
minimize ['minimaiz]	vt. 使减到最少，最小化
detrimental [,detrɪ'ment(ə)l]	adj. 不利的，有害的；n. 有害的人
aforementioned [əfɔː'menʃənd]	adj. 上述的，前面提及的
thereby [ðeə'baɪ]	adv. 从而，因此，在那附近，在那方面
adaptive [ə'dæptɪv]	adj. 适应的，适合的
adaptive control	自适应控制
implementation [ɪmplɪmen'teɪʃ(ə)n]	n. [计] 实现，履行，安装启用

➪ Notes

[1] Likewise, material joining is conventionally accomplished with thermal energy sources such as burning gases and electric arcs.

参考译文：同样，材料连接通常是通过诸如燃烧气体和电弧的热能源来完成的。

burning gases and electric arcs 意为"燃烧的气体和电弧"。

[2] Material removal can now be accomplished with electrochemical reactions, high-temperature plasmas, and high-velocity jets of liquids and abrasives.

参考译文：现在材料的去除可以通过电化学反应、高温等离子、高速液体和磨料射流等完成。

electrochemical reaction, high-temperature, and high-velocity jets of liquids and abrasives 意为"电化学反应、高温等离子、高速液体和磨料射流"。

[3] In other cases, nontraditional processes are used to reduce the number of rejects…

参考译文：在其他情况下，采用特种加工工艺可以减少不合格品的数量……。

number of reject 意为"不合格品的数量"。

[4] Additionally, computer control assures reliability and repeatability…

参考译文：此外，计算机控制保证了可靠性和重复性。

repeatability 意为"重复性"。

⊠ Questions

1. Can you summarize the history of tools material evolving?
2. Why are manufacturers of today faced with tough questions?
3. When did nontraditional manufacturing processes begin in history?
4. What is the difference between conventional and nontraditional manufacturing processes?
5. Why will nontraditional manufacturing processes increasingly develop in the future?

Reading Material: Cutting Forces and Cutting Power

Cutting Forces

To start the cutting process it is necessary to apply a force sufficiently great and suitably situated in space. This force must be able to overcome the resistance of the material to elastic and plastic deformation, the frictional resistance and the resistance to the destruction of cohesion and to the creation of a new material surface.

Such a force is called the cutting force and the entire resultant of the machined material is called the cutting resistance. It is represented by a vector force of a magnitude equal to that of the cutting force and acting in the same direction but in the opposite sense. Thus the cutting force is the force exerted by the tool point upon the material being machined and the cutting resistance the force exerted by the material upon the tool point.

The magnitude and direction of both the cutting resistance and cutting force change in space depending on many factors involved in the cutting process.

In order to determine the cutting force unequivocally, its components are defined in an orthogonal reference system.

The system involving the tool consists of:

(1) the straight line z, parallel to the direction of the primary motion. The projection of the cutting force on this line is called the tangential or primary cutting force Fv;

(2) the straight line y, parallel to the reference plane and perpendicular to the vector of primary motion velocity. The projection of the cutting force on this line is called the thrust force Fp;

(3) the straight line x, perpendicular to both the former ones.

(4) The projection of the cutting force on this line is called the axial cutting force F_f. This component is also called the feed force, if its direction coincides with that of the feed motion. Components in the x-y-z system directly characterize the cutting process itself.

The cutting force is a resultant of force exerted upon the rake and flank. This can be explained in a simplified way by means of Figure 20-1.

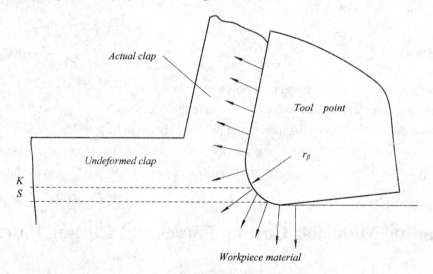

Figure 20-1　Diagram of the division into zones of action of face and flank

The actual cutting edge always has a nose radius $r_\beta > 0$. Considering the simplified model of chip formation by orthogonal cutting, it may be assumed that the imaginary shear surface will start from a certain point K, situated on the curved part of the tool point profile. Thus it may be assumed that point K divides the zones of action of the rake and flank.

Under this assumption the forces exerted by the rake will mainly do the chip formation work and the frictional work of the chip against the rake face.

Unit 20 Nontraditional Manufacturing Processes

Forces exerted upon the flank are mainly carrying out the work of elastic and plastic deformation of the layer situated below the K-k line, as well as the frictional work of the machined surface against the flank.

The equations for component cutting forces may be written as follows:

$$F_v = hbk_8 \tag{20-1}$$
$$F_p = w_y hbk_8 \tag{20-2}$$
$$F_f = w_x hbk_8 \tag{20-3}$$

K_8 is called the specific cutting resistance or cutting stress. In other words the cutting stress or specific cutting resistance K_8, is the quotient of the main cutting force and the cross-sectional area of the undeformed chip.

In order to calculate the magnitude of the cutting force components according to eqn. stated above the values of K_8, w_y and w_x are determined in an empirical and statistical way. The dependence of these magnitudes upon a number of factors, involving the machine tool, tools, material and medium, is represented graphically and then suitable equations are chosen (approximated), which define those relationships with sufficient accuracy.

The empirical and statistical equations do not always reflect the physical side of the cutting process in a correct way. However, they furnish useful practical relationships between the cutting resistance and various cutting conditions.

For the user's convenience nearly all the relationships are approximated by products of power functions whose bases are constituted by cutting parameters, and the exponents, consistent within certain ranges, characterize the intensity of influence of the given parameter upon the cutting resistance.

The approximation by a power function may in certain cases by the cause of difficulties as far as true presentation of the actual influence of the given parameter is concerned. It is due to the fact that in some equations the so-called constant coefficients and exponents may in actual fact vary.

Therefore, when empirical and statistical equations based on such functions are used, it should be remembered that:

—they may be applied only within a range which has been studied previously; extrapolation of empirical and statistical equations is not permissible because of the risk of introducing errors of unknown magnitude;

—they are built on the assumption that the particular factors are independent of one other; actually, there exists an indirect or direct interdependence of those factors, and therefore the equations based on statistical means yield only approximate values;

—out of the many cutting conditions, only those whose influence is most important or which vary most frequently are shown separately in the equation; the influence of the remaining conditions is taken account of in the form of so-called constants or correction factors, the values of which are strictly established only for a certain range of machining processes.

For example, the empirical and statistical method was used to determine the equation for specific cutting force per unit cross-sectional area of the undeformed chip as

$$K = \frac{C_z}{h_z^m b_z^n} \quad (20\text{-}4)$$

Where, C_z —coefficient taking care of the influence of all factors not accounted for separately in the equation; mz and n_z— exponents established in an empirical way for the given range.

In view of this relationship, the equations for the cutting force components may be rewritten in the form

$$F_v = C_z h_z^e b_z^u \quad (20\text{-}5)$$

$$F_p = C_y b_y^e h_y^u \quad (20\text{-}6)$$

$$F_f = C_z b_z^e h_z^e \quad (20\text{-}7)$$

In eqn: $C_y = w_y C_z$; $C_x = w_x C_z$; and $u_z = 1-m_z$, $u_y = 1-m_y$; $u_x = 1-m_z$; $e_z = 1-n_z$; $e_y = 1-n_y$; $e_x = 1-n_x$.

Particular values of the coefficients C_x, C_y and C_z as well as of exponents U_x, U_y, U_z, e_x, e_y and e_z depend, among other factors, on the kinematics of cutting.

Therefore, these values will be given when the cutting methods and their varieties are discussed.

Cutting Power

Generally, cutting power is defined as $N_c = \dfrac{dE}{dt}$

Where t is cutting time.

In numerous cases of cutting, N_c is constant, e.g. when the speeds of basic motions are constant and the cutting force varies so slightly that its mean value may be considered as constant. Then

$$N_c = \frac{E}{t} \quad (20\text{-}8)$$

The total cutting energy consists of the all components of the cutting force. Thus, if the cutting power in kilowatts is to be determined, then

$$N_c = \frac{F_v V_m}{6120} + \frac{F_p V_p}{6120} + \frac{F_f V_f}{6120} \quad (20\text{-}9)$$

Where: F_v, F_p and F_f mean values of cutting force components; V_m, V_p and V_f mean values of speeds in the direction of action the corresponding components.

Denominator value is the product 60×75×1.35 when speeds are given in m/min and forces in kg.

The equation, used for the calculation of power needed in the particular cutting methods, is mostly reduced to one or two terms, since the rest may be omitted as equal to zero or negligibly small.

Mechanical energy is partly transformed in the cutting process into heat and partly used to increase the potential energy of the deformed crystal lattice

However, it follows from the results of Rebinder and Epifanov that the part of energy used to increase the potential energy of crystal lattices is negligibly small (less than 1%) and thus it can be assumed that practically all mechanical energy is turned into heat.

The amount of heat generated in the cutting process may therefore be calculated as equivalent to the cutting energy and the rate of heat emission may be defined by

$$Q = \frac{F_v V_m}{427} \text{cal/min} \tag{20-10}$$

where 427 kg m/cal is the mechanical equivalent of heat.

The heat emitted is absorbed by the chip (Q_c), and by the tool (Q_s) transmitted to workpiece. Material (Q_w) is absorbed by the cutting fluid and ambient air (Q_a). Rediation is omitted as negligible. This can is be written as

$$Q = Q_a + Q_c + Q_w \tag{20-11}$$

A knowledge of heat distribution among chip, tool point, material being machined and the cutting fluid and ambient air is of great practical importance. Heat distribution is dependent on cutting conditions kind of chip, cooling method, properties of workpiece material and on tool point material and shape.

On the basis of studies conducted so far, the following conclusions may be drawn:

(1) One of the factors having most important influence on heat distribution is cutting speed.

(2) Increase of cutting speed causes a reduction of the heat transmitted into the material.

From a practical point of view, however, the important factor is not the amount of heat absorbed by the tool point and material being machined, but, in the first place, the temperature level and distribution.

◇ New Words and Expressions

cohesion [kəʊˈhiːʒn]	n. 黏附，黏结
crystal attice	晶格
destruction [dɪˈstrʌkʃn]	n. 破裂
power function	幂函数
in reality	实际上
in view of	考虑到，由于
out of	从……当中，来自
take account of	考虑，注意
correction factor	修正系数

on the basis of 根据，以……为基础
cutting fluid 切削液
percentage share 百分比

⊠ Questions

1. What is cutting force?
2. What does the orthogonal reference system consist of?
3. How are the equations for component cutting forces represented?
4. What is cutting power?
5. Can you explain the equation $N_c = \dfrac{F_v V_m}{6120} + \dfrac{F_p V_p}{6120} + \dfrac{F_f V_f}{6120}$?

PART 5　Engineering Machines

Unit 21　The Two Luffing Cable Cranes

General Description

The two luffing cable cranes are destined for the construction of the Three Gorges Dam in the valley of the Yangtze River.

The jobsite is situated near Sandouping, Yichang in Hubei province in the People's Republic of China.

The cable cranes, representing the main transport facility of a dam construction jobsite, have been designed for the purpose of transporting concrete, formwork items, concrete pre-fab items, reinforcing material, building machinery and machine components such as penstocks, generators, lock gates, etc. The cable cranes that will be used on the jobsite are of the luffing type, i.e. it has been given that name because for moving the track rope, the towers or masts have to luff to the upstream or downstream direction.

Both the slewing masts of each single crane between which the track ropes are installed, are situated at an elevation of +185m on block 8 already concreted on the left bank and on the longitudinal coffer dam at the elevation of +160 m[1].

Each mast has a height of 125 + 25 = 150m, and is firmly anchored by rear stay ropes in longitudinal direction. On their sides, the masts are held respectively moved by 6-strand luffing ropes actuated by winches[2].

The spans are:

Rear stay ropes on the left bank approx. 270m,

Rear stay ropes on the right bank approx. 604.5m,

Track ropes: approx. 1416m,

The machine houses for the winches of the cranes are arranged on the foot of the masts.

In the machine houses on the foot of the head towers are arranged the winches for hoisting crab traversing and mast luffing.

In the machine houses on the foot of the tail towers are only situated the winches for mast luffing. This machine houses are therefore called luffing winch house.

The control electronics are each located in separate rooms, for thus avoiding any problems as might occur by dust or also temperature variations[3].

There are three possibilities to control the winches.

Normally the crane is controlled by the driver's cabin that is movable between the machine side and the counter side on a long cable.

The second way is to control the cable crane by a wireless portable control unit. For each crane there is one mobile unit.

The third possibility to control the winches is the local control device nearby the machines for the hoist, crab traversing and luffing systems. The local control device is for purpose of easy adjustment and repairing.

Because of the local situation, there are certain areas of the working field with obstacles. These areas are locked by the programmable computer control system.

For the control systems also see the Electrical documents of ABB Company.

The crab traversing winch is driven by a DC-motor that allows for an infinitely variable speed regulation. For physics reasons the acceleration shouldn't be much beyond 0.75 m/s^2. Thereof result acceleration and braking times of approx. 10 seconds (braking torque about 6000 Nm) [4]; respective path is approx. 37.5m. With commissioning operations the mechanical brake shall be set in a way that the loaded crab will come to be stopped after at the latest 12 seconds (braking torque approx. 6000 Nm). Here one has to check periodically for ensuring a uniform running in the two grooves (see Section "Maintenance"). Between the two grooves the crab traversing rope is led over the return pulley, at the shaft of which is also fitted the screw-type limit switch and the position indicator, just also the tachometers for measuring the speed and for control of a possible over speed. Although the screw-type limit switch and the position indicator are driven by the non-driven return pulley, it is unavoidable that after a duration of operation occurs a certain miss-indication, so that the correction will have to be made (see section "Traversing Limit Switches and Position Indicators").

For safety reasons have been installed, however, at the first and at the last rope carrier additional limit switches for monitoring the crab speed. Since these limit switches cannot shift out of their set position, the signal is absolute. Finally the crab traversing is then stopped by a lever limit switch attached at each end of the track rope.

The tachometer for monitoring a possibly occurring over speed disconnects the crab traversing gear at 115% rated speed by "emergency off".

For providing the operation and maintenance team with information concerning the plant output, the crab traversing gear is equipped with an operation hour's counter and a cycle counter. Here the operation hours counter is detecting the time in which the crab traversing motor is working, and the cycle counter the number of traversing runs. Therefore, the cycle counter is set in a way that after each motor operation time of about 10 seconds in one direction another half cycle is counted up [5]. In case the travel is interrupted and continued, thereafter in the same direction no new half cycle will be added. In case the direction of travel is reversed, however, and then will result after another 10 seconds a new half cycle [6].

The cycle counter shall be a help for the maintenance team to get a better idea about the working life of various wearing items such as trolley traversing rope, hoisting rope, rope carriers, drive pulley lining, etc.. It is necessary that with the exchange of these items are always read at the cycle counter the resp. working cycles and that the results will be registered.

As an additional maintenance help a lubricants tank is mounted to the traversing winch for lubrication of the crab traversing rope.

Traversing Limit Switches and Position Indicators

As mentioned in the previous section, the position indicator and the traversing limit switch (a screw-type limit switch) are arranged on the shaft of the return pulley, from which they can both be separated by a magnetic tooth coupling. With erection the traversing limit switch is set in a way that the crab stops travel 130m before reaching the machine side [7], or resp 110m before the counter side. The position indicator should then show the corresponding value.

During operation, the crab is automatically braked down electrically to $v = 0$ m/s, when one of the two end positions is reached. On the remaining track rope length travel is then only possible, when the by-pass button is depressed. By depressing the by-pass button the travel speed is also limited to $v < 20\%$ of v rated. Running in opposite direction within this range, there is no speed limitation.

In order to ensure principally a reduction of the crab speed, the plant has been equipped with two additional stationary monitoring limit switches. These are arranged on always the first and the last rope carrier. As these are in parallel connection with the screw-type limit switch, they also cause a reduction of the crab traversing speed to $v = 0$ m/s by electrical breaking. To continue the travel to the end of the track rope is only possible by depressing the by-pass button.

The crab traversing is finally stopped electrically by each one limit switch at the end of the track rope.

For being capable on principle to take into consideration further obstacles in the travel area, the screw-type limit switch has been equipped with further contacts.

By the unavoidable slip between return pulley and crab traversing rope, and by the retensioning of the crab traversing rope, a correction of the screw-type limit switch and the position indicator will become necessary. This correction should be made at least once a day with start of the new shift. For this purpose the crab must run to a position that is known to the crane driver and from which is known the exact distance up to the respective end of the track rope (as reference position would be suitable a rope carrier or the concrete platform) [8]. In case, at that position of the crab, the value shown for the position of the crab is not in accordance with the actual distance, then the screw-type limit switch and the position indicator must be corrected.

For this purpose the crab must be run into a position, where the position indicator is showing the set distance (e.g. at the first rope carrier 50 m). As soon as this figure is appearing on the display, crab traversing must be interrupted and the synchronous button will separate the return pulley from the screw-type limit switch and the position indicator will limit the crab traversing speed to $v < 20\%$ of v rated. The synchronous button must be depressed during the whole synchronization procedure. The crab then runs to the reference point that has been set to the display (e.g. the first rope carrier).

Maintenance

Beside the general maintenance and inspection work, and as is further described under the Section "General Maintenance Instruction" and "Lubrication", the owner should take care of a uniform running of the ropes in the Becorit lining. As soon as irregularities occur, these can be eliminated by a correction of the alignment of the drive pulley and of the return pulley. For this purpose setting screws are fitted to the pillow blocks of the drive pulley and of the return pulley. By the clamping set incorporated in the return pulley, the latter can also be shifted axially.

The drive pulley of the crab-traversing winch is designed as 2-groove drive pulley lined with Becorit. Depending on the varying rope tension of the crab traversing rope in the two rope grooves and the different rope slip there results a different wearing of Becorit lining in the two grooves[9]. A different groove. depth means also a different circumferential speed and this results forcibly in an increased rope slip and therefore in a heavier wearing of the lining.

Therefore the drive pulley grooves must be checked frequently for their depth (by means of gauge). With a depth difference of 2 mm one groove is to be turned until the two grooves have again the same depth[10]. Turning speed for cutting is 2 to 4 m/sec.

The Becorit lining should be replaced as soon as the remaining thickness has come down to approx. 20 mm. This corresponds to a total groove depth of 40 mm at a lining thickness of 60 mm.

◆ *New Words and Expressions*

luff [lʌf]	n.	俯仰
penstock ['penstɒk]	n.	压力水管
mast [mɑːst]	n.	桅杆，立柱
upstream [ˌʌp'striːm]	n.	上游
winche [wɪntʃ]	n.	绞车
pulley ['pʊli]	n.	皮带轮，滑轮
maintenance ['meɪntənəns]	n.	维护，保养
cable crane	n.	缆机
concrete pre-fab item	n.	混凝土预制件
reinforcing material	n.	钢筋，加强材料
lock gate		闸门
track rope		跑索
coffer dam		围堰
at an elevation of		在……的高程，在标高为……
be arranged in		布置在……，安装在……
be located in		位于……
allow for		考虑到，容许
in one direction		朝一个方向

be equipped with	装配有，配备有
in parallel connection with	与……并联
limit switch	限位开关

⇨ Notes

[1] Both the slewing masts of each single crane between which the track ropes are installed, are situated at an elevation of +185 m on block 8 already concreted on the left bank and on the longitudinal coffer dam at the elevation of +160 m.

参考译文：缆机的两个转动起重塔杆，中间安装跑索，布置在左岸已浇筑的高程为 185 m 的 8 号坝块上和高程为 160m 的纵向围堰上。

between which the track ropes are installed 是定语从句，其中 which 代表前面的 masts。

[2] On their sides, the masts are held respectively moved by 6-strand luffing ropes actuated by winches.

参考译文：在两侧，起重塔杆通过绞盘绞动的六股俯仰索加以固定，并控制俯仰动作。

moved 引导的词组作状语，修饰前面 held 所表示的动作；其中 actuated 作定语，修饰 ropes。

[3] The control electronics are each located in separate rooms, for thus avoiding any problem as might occur by dust or also temperature variations.

参考译文：电子控制设备分别安装在单独的机房里，因为这样能避免由于灰尘或温度变化可能引起的问题。

as 引导的是定语从句，修饰 problem。

[4] Thereof result acceleration and braking times of approx. 10 seconds (braking torque about 6000 Nm).

参考译文：由此产生的加速度和制动时间约为 10 s (制动转矩约为 6000 Nm)。

由于主语太长，故用倒装句；句子的谓语是 result，其后的词组为主语；thereof 是谓语的状语，意为 from that。

[5] Therefore the cycle counter is set in a way that after each motor operation time of about 10 seconds in one direction another half cycle is counted up.

参考译文：因此，行程计数器设置为马达在同一方向每运行约 10 s 后，则将另一半行程计算在内为一次行程。

[6] In case the travel is interrupted and continued, thereafter in the same direction no new half cycle will be added. In case the direction of travel is reversed, however, then will result after another 10 seconds a new half cycle.

参考译文：如果中断并且随后又继续朝同一方向横移时，则仍计为一次行程；然而，如果横移的方向相反时，则再过 10 s 后就另计一次行程。

[7] With erection the traversing limit switch is set in a way that the crab stops travel 130 m before reaching the machine side…

参考译文：安装时对限位开关进行设定，以保证缆车能在到达尽头(机器边缘)前 130 m

时停止。

　　with 表示"与……一起"。如后面跟着表示动作意义的词时，则理解为"在……的同时"。本文中另一处也有这种用法："It is necessary that <u>with</u> the exchange of these item are read at the cycle counter the respective working cycles and that the results will be registered."。

　　[8]　For this purpose, the crab must run to a position that is known to the crane driver and from which is known the exact distance up to the respective end of the track rope (as reference position would be suitable a rope carrier or the concrete platform).

　　参考译文：为此目的，必须将缆车行进到缆机驾驶员熟悉的位置，并且知道从这个位置至跑索两端的准确距离(某个缆索架或混凝土平台可能会成为一个合适的参照位置)。

　　which 引导的从句和 known to the crane driver 是并列的，都修饰 position。括号中是一个倒装句，正常语序为：a rope carrier or the concrete platform would be suitable as reference position。

　　[9]　Depending on the varying rope tension of the crab traversing rope in the two grooves and the different rope slip there results a different wearing of Becorit lining in the two grooves.

　　参考译文：由于在两个缆索槽中缆车横移缆索的张力不同和缆索滑动的不同，结果会导致两个槽内的比克瑞特内套出现不同程度的磨损。

　　results 前的 there 是引导词，与 there is a book on the table 中的 there 一样，只起引导作用，没有意义。当主语太长，而谓语又是不及物动词时，常用 there 作句子的引导词，如：There appeared a man who was dressed in a long, black robe.

　　[10]　With a depth difference of 2 mm one groove is to be turned until the two grooves have again the same depth.

　　参考译文：当两个槽的深度相差 2 mm 时，则要车削其中一个槽，直至两个槽的深度相同。

◇ Questions

1. What is the purpose of designing cable cranes?
2. In what way can the winches be controlled?
3. Why must the drive pulley grooves be checked frequently for their depth?
4. For what purpose is the plant equipped with two additional stationary monitoring limit switches?
5. How many possibilities to control the winches are there?

Reading Material: Hydraulic System

　　The history of hydraulic power is a long one, dating from man's prehistoric efforts to harness the energy in the world around him [1]. The only sources readily available were the water and the wind-two free and moving streams.

The watermill, the first hydraulic motor, was an early invention. One is pictured on a mosaic at the Great Palace in Byzantium, dating from the early fifth century. The mill had been built by the Romans. But the first record of a watermill goes back even further, to around 100 BC, and the origins may indeed have been much earlier. The domestication of grain began some 5000 years before and some enterprising farmer is bound to have become tired of pounding or grinding the grain by hand. Perhaps, in fact, the inventor was some farmer's wife, since she often drew the heavy jobs.

Many mills stayed in use until the end of the 19th century, but they had been in the process of being replaced by the steam engine as a source of power for the previous 150 years. The transmission of the power generated by the waterwheels was by shafting and crude gears or pulleys. The steam engine for the first time employed an enclosed moving stream of fluid (steam) under pressure—the principle to be later employed in hydraulic power transmission for transmitting power continuously from the point of generation to the place where it was used.

The transmission of hydrostatic power to a distant point began long before that, however, Hero of Alexandria, in the first century AD, built a device in which a fire on the temple altar expanded air in a closed container. The air pressure forced water to travel along a tube to the temple doors where, spilling into a container, it provided the force through ropes and pulleys to magically open the doors.

Hero also produced a steam engine of sorts. Its motion depended on the reaction forces from jets of steam, as in the present steam turbine, but it was little understood and it was no more than an interesting curiosity.

The invention of early forms of the hydraulic pump has a similarly early origin. The first pumps were not used to develop hydraulic power, but only to transfer water for irrigation or to remove it from mines.

Archimedes applied the principle of the screw to hydraulic machines in the third century BC. His screw pumps were used to raise water for irrigation or to the level of aqueducts. The piston pump, the first mechanical device capable of generating pressure in a column of liquid, is believed to have come from Egypt at a similarly early date[2]. Like the screw pump, it was used only as a means for moving water and not as a means for generating hydraulic power.

In more recent years, the role of leadership in hydraulic power application has been taken over largely by some of the larger earthmoving and construction equipment manufacturers[3]. The total power involved is often greater than that required in even the largest aircraft systems. The concentration of this power in a few very large loads, as compared to the multitude of smaller loads in an airplane, has spurred the system designer to find new, creative ways to distribute and control the power. This designer has succeeded in doing to obtain maximum precision of control and productivity while minimizing power consumption.

In modern mobile and industrial applications, the system designer has made full use of the wonderful flexibility and adaptability of hydraulics. It can readily be directed around corners and past involving joints that virtually defy mechanical power transmission. It can be divided easily to serve individual loads or recombined to power a high, single-load demand. Electrical and pneumatic power distribution have similar flexibility, but lack the compactness and certain of the other important characteristics of hydraulics.

A very significant feature of hydraulic power is its extreme power density [4]. Figure21-1 shows a comparison of a 400 hp (1hp=0.7457kW) hydraulic pumping source with a diesel engine and an electrical motor of the same power capacity. The contrast is obvious and is one that offers both an opportunity to the user of hydraulic power and a challenge to the designer of hydraulic components.

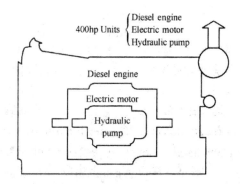

Figure 21-1　Comparative size of diesel, electric and hydraulic power units

The high-power density in a hydraulic pump or motor creates design challenges beyond those found in many other products. High loads on rapidly sliding surfaces, extreme fluid velocities and rapidly recurring pressure applications on pumping elements (vanes, pistons, gears) demand careful design attention to assure long life, complete reliability and efficient operation. The user, on the other hand, reaps the benefit of a family of apparatus that is highly responsive, finely controllable and simple to apply. When the designer understands the user's needs and the user understands the equipment's full capabilities, the maximum potential of hydraulic power can be achieved.

◇ *New Words and Expressions*

prehistoric [ˌpriːhɪˈstɒrɪk]	adj. 史前的，很久以前的
harness [ˈhɑːnɪs]	vt. 利用 (风等) 作动力，治理，控制
watermill [ˈwɔːtəmɪl]	n. 水车，水磨
mosaic [məʊˈzeɪɪk]	n. 镶嵌细工，马赛克
domestication [dəˌmestɪˈkeɪʃn]	n. 家养，驯养
waterwheel [ˈwɔːtəwiːl]	n. 水轮，水车，毂辘
pulley [ˈpʊli]	n. 滑车，滑轮
altar [ˈɔːltə(r)]	n. 祭坛，圣坛
spill [spɪl]	n. 溢出，流出
aqueduct [ˈækwɪdʌkt]	n. 高架渠，渡槽

earthmoving ['ɜːθmuːvɪŋ]　　　　　　adj. 大量掘土的，大量运土的
spur [spɜː(r)]　　　　　　　　　　　v. & n. 刺激，激励，鼓励，推动
flexibility [ˌfleksə'bɪlətɪ]　　　　　　n. 柔韧性，柔性
adaptability [əˌdæptə'bɪlətɪ]　　　　n. 适应性
defy [dɪ'faɪ]　　　　　　　　　　　　v. & n. 使不能 (难以，落空)，向……挑战
compactness [kəm'pæktnɪs]　　　　n. 紧密，紧密度，简洁
diesel ['diːzl]　　　　　　　　　　　n. 柴油机

⇨ Notes

[1]　The history of hydraulic power is a long one，dating from man's prehistoric efforts to harness the energy in the world around him.

参考译文：水力的历史由来已久，始于人类为利用它周围的能源而做出的努力。

dating from…是现在分词作状语。

[2]　His screw pumps were used to raise water for irrigation or to the level of aqueducts. The piston pump，the first mechanical device capable of generating pressure in a column of liquid，is believed to have come from Egypt at a similarly early date.

参考译文：柱塞泵是历史上第一台能使液柱内产生压力的机械装置，它被认为来源于埃及且与此(螺旋泵)有着同样悠久的历史。

句中的 the first mechanical device capable of generating pressure in a column of liquid 作 the piston pump 的同位语。

[3]　In more recent years，the role of leadership in hydraulic power application has been taken over largely by some of the larger earthmoving and construction equipment manufacturers.

参考译文：近年来，一些规模较大的生产土建设备的厂商在液压动力应用方面一直占着主导地位。

这是一个被动语态的句子，主语为 the role of leadership in hydraulic power application，其中 in hydraulic power application 为介词短语作 the role of leadership 的定语；in more recent years 是介词短语，作全句的状语。

[4]　A very significant feature of hydraulic power is its extreme power density.

参考译文：水力的一个明显的特征就是它具有极高的功率密度。

句子中的 power density 可以翻译为 "功率密度"。

⊠ Questions

1. Can you describe the history of the watermill?
2. What are the significant features of hydraulic power?
3. What is the contribution of Hero of Alexandria?
4. When were mills replaced by the steam engine?
5. What significant feature does hydraulic power have?

Unit 22　Komatsu D375A Bulldozer

Outstanding Productivity

Stress-absorbing Undercarriages Offer Mighty Traction

The sprockets and idlers are mounted through elastic supports, softening shocks to both the machine and the operator, caused when the machine front makes ground contact, following travel over hard or rocky terrain [1]. Since track rollers are rigidly mounted on the track frames, they do not oscillate and they allow track shoe grousers to bite the ground with near-to-constant force, increasing traction and reducing shoe slippage [2]. Thus, the machine's force is effectively converted into traction.

A High-potency Engine for Powerful Dozing and Ripping

The SA6D170 engine delivers a dynamic power of 525 HP (391 kW) at 1800 RPM and a tenacious torque of 267 kg-m (1,931 ft-lb/2,618Nm). This, together with the heaviest machine weight produces the strongest drawbar pull of 51.5 tons (when shoe slipping with F2 gear), making the D375A a superior bulldozer in both ripping and dozing production [3].

High Blade and Ripper Performance

A large blade capacity of 17.3 m^3 (semi-U dozer) and 20.9 m^3(U-dozer) means increased production (Figure 22-1). High-tensile-strength steel comprising the front and side of the blade increases durability. The variable giant ripper features long sprocket center-to-ripper point distance, making ripping operation easy and effective along with great penetration force (Figure 22-2)[4].

Figure 22-1　Komatsu D375A bulldozer

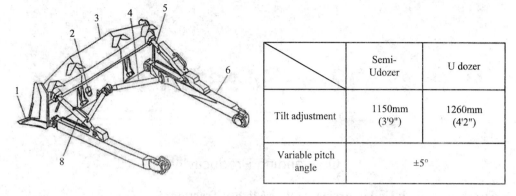

	Semi-Udozer	U dozer
Tilt adjustment	1150mm (3'9")	1260mm (4'2")
Variable pitch angle	±5°	

Figure 22-2　Large blade capacity

Breakthroughs in Fuel Economy

Automatic lockup system is effective for dozing loose soil. When more power is needed, especially in long-distance dozing or hauling, turn on the LOCKUP switch and the lockup clutch in the torque converter is automatically engaged or disengaged (in other words, direct drive or torque converter drive is automatically selected) depending on the revolutions of the torque converter output shaft. The result: more effective use of engine power with less fuel consumption.

The Komatsu SA6D170 Engine Realized Minimum Fuel Consumption

This newly designed engine features a direct-injection system, turbocharger and after cooler to maximize fuel efficiency.

The engine is mounted on the main frame through rubber cushions to minimize noise and vibration. For further convenience, fuel adjustment is unnecessary up to an altitude of 3000m (9,840ft).

The Dual Tilt Dozer (option) reduces operator effort while increasing productivity.

1) Optimum blade cutting angle for all types of materials and ground inclinations can be selected on the go for increased loads and increased production.

2) Digging, hauling and dumping are easier and smoother with less operator fatigue.

3) Increased dozer tilt angle and tilt speed by twice that of a conventional single tilt system[5].

Minimum Downtime

Stress-absorbing Undercarriages Extend Service Life. Rubber cushions surround the sprocket and idler shafts to soften shocks caused when the machine goes over obstacles, thus significantly reducing operator fatigue and extending component life.

Unique Modular Design Makes it Easy to Remove Power-train Components without Oil Spillage. The sealed, modular design allows the power-train components to be mounted/dismounted without any oil spillage, making servicing work clean, smooth and easy.

Advanced Monitoring System Helps to Prevent Minor Problems from Becoming Major Ones. Conditions of both check-before-starting items and caution items appear on the liquid crystal monitoring panel, preventing the development of serious problems (Figure 22-3). The continuous condition check allows the operator to concentrate his attention on the controls.

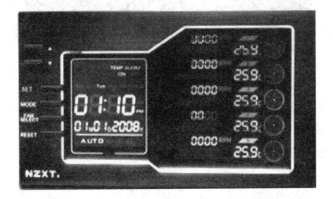

Figure 22-3 The liquid crystal monitoring panel

Tilt-back ROPS frame. The entire ROPS structure and cab are tilted backward, facilitating accessibility for the mounting/dismounting of power-train components [6].

Centralized Service Station for Simplified Maintenance. To assure convenient maintenance, the transmission and torque converter oil filters are arranged side by side next to the power-train oil level gauge. In addition, oil pressure checking ports for power-train components are centralized on the service station, promoting quick and simple upkeep.

Other Maintenance-aid Features. The tilt-type radiator allows ready cleaning of the radiator cores. Hydraulic piping for the blade tilt cylinder is completely housed in the dozer frame ensuring damage protection from external objects.

Operator Comfort

An Extra-low Machine Profile Means Excellent Dynamic Stability. The innovative low-profile design assures excellent machine balance and a low center of gravity, making the D375A dynamically stable and controllable, accounting for greater operator confidence and comfort.

Large Cab Window Area Offers Panoramic View. The novel cab design and immense glass area provide a wide visual range. In addition to the rubber cushioned floor frame mount, the cab has built-in sound-absorbing urethane foam and is entirely sealed, resulting in minimal noise and vibration disturbances. To provide ample legroom, the instruments are concentrated on the right side of the operator's seat. The highly responsive hydraulic system offers ease of fine controlling, enabling precise dozing operation. The rotatable seat offers optimum visual range (Figure 22-4).

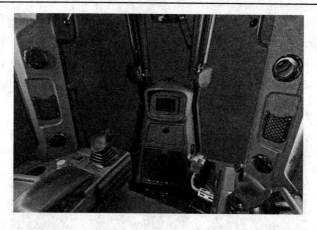

Figure 22-4　Optional equipment is included in the cab

◇ New Words and Expressions

traction ['trækʃn]	n. 抓地力
sprocket ['sprɒkɪt]	n. 链轮
idler ['aɪdlər]	n. 惰轮
bulldozer ['bʊldəʊzə(r)]	n. 推土机
ripper ['rɪpər]	n. 挖掘头
breakthrough ['breɪkθruː]	n. 突破
cushion ['kʊʃən]	n. 垫套，缓冲垫
downtime ['daʊntaɪm]	n. 停机时间
panoramic [ˌpænə'ræmɪk]	adj. 全景的
novel ['nɒvl]	adj. 新奇的
rotatable ['rəʊteɪtəbəl]	adj. 可旋转的
stress-absorbing	减压
dual tilt	双倾斜(缸)
on the go	在繁忙中
account for	说明……的原因

⇨ Notes

[1]　The sprockets and idlers are mounted through elastic supports, softening shocks to both the machine and the operator, caused when the machine front makes ground contact, following travel over hard and rocky terrain.

参考译文：链齿轮和惰轮安装在弹性支座上从而减轻推土机驶过坚硬或岩石路面，机前端接触地面时对操作员和机器造成的震动。

softening 引导的分词短语作状语，表示前面句子中的冲击动作结果；caused 引导的分词短语作定语，修饰前面 shocks 一词；following 引导的分词短语作状语，表示 when 引导的从句中的动作发生的时间，意为"在……之后"。

[2] …and they allow track shoe grousers to bite the ground with near-to-constant force, increasing traction and reducing shoe slippage.

参考译文：(它们)可使履带靴片以接近恒定的力量抓地，从而增加抓地力，减少打滑。

在英语科技文章中，在主句后用现在分词短语来表示主句中的动作所产生的实际效果。因此在汉语译文中，常在分词前添加"因此"、"从而"等类似的词。Note [1]中的 softening 也是如此。现在分词的此种用法，与下面例句中的进行时十分相似：When a child cries, he is releasing his anxiety. (小孩哭实际上是在缓解他的焦虑)。

[3] This, together with the heaviest machine weight produces the strongest drawbar pull of 51.5 tons (when shoe slipping with F2 gear) …

参考译文：所有这些，再加上机器的巨大(最重)的重量，就产生出最强的拉杆牵引力，达到 51.5 吨(在前进二挡打滑时)，……。

when shoe slipping with F2 gear 是省略句，其中的 shoe slipping 并不是主谓关系，而是名词词组。

[4] The variable giant ripper features long sprocket center-to-ripper point distance, making ripping operation easy and effective along with great penetration force.

参考译文：可调节的巨型挖掘头的特点是从链齿轮中心至挖掘头点之间的距离长，使挖掘头作业时的穿透力强，作业轻松有效。

features 是动词单数第三人称，动词的意思为"突出……的特点"。本单元中另有一处也有 feature 作动词的用法：This newly designed engine features a direct-injection system, turbocharger and after cooler to maximize fuel efficiency. 这种新设计的发动机的突出特点是配备直喷系统、涡轮增压和后冷却系统，最大程度地提高燃料的效率。

[5] Increased dozer tilt angle and tilt speed by twice that of a conventional single tilt system.

参考译文：与常规的单倾斜缸系统相比，双倾斜缸推土铲的倾斜角度和倾斜速度增加了一倍。

本句可改写成 The dozer tilt angle and tilt speed are increased by twice that of a conventional single tilt system, 这样就更容易理解了。by 引导一个数量词，表示差额关系。另如：The budget was cut by 5%.(预算被减少了百分之五)，by 5% 表示与原来的预算相差百分之五。

[6] The entire ROPS structure and cab are tilted backward, facilitating accessibility for the mounting/dismounting of power-train components.

参考译文：将整个 ROPS 结构和驾驶室向后倾斜，可便于动力总成部件的安装与拆卸。

⊠ Questions

1. What is the function of the SA6D170 engine?
2. Please describe the working principle of automatic lockup system?
3. Why can the Dual Tilt Dozer reduce operator effort while increasing productivity?

4. How can advanced monitoring system help to prevent minor problems from becoming major ones?
5. What comfort can Komatsu D375A Bulldozer bring to operators?

Reading Material: Lubrication

Although one of the main purposes of lubrication is reduce friction, any substance-liquid, solid, or gaseous—capable of controlling friction and wear between sliding surfaces can be classed as a lubricant.

Varieties of Lubrication

Unlubricated Sliding. Metals that have been carefully treated to remove all foreign materials seize and weld to one another when slid together. In the absence of such a high degree of cleanliness, adsorbed gases, water vapor, oxides, and contaminants reduce friction and the tendency to seize but usually result in severe wear; this is called "unlubricated" or dry sliding.

Fluid-film Lubrication. Interposing a fluid film that completely separates the sliding surfaces results in fluid-film lubrication. The fluid may be introduced intentionally as the oil in the main bearing of an automobile, or unintentionally, as in the case of water between a smooth rubber tire and a wet pavement. Although the fluid is usually a liquid such as oil, water, and a wide range of other materials, it may also be a gas. The gas most commonly employed is air.

To keep the parts separated, it is necessary that the pressure within the lubrication film balance the load on the sliding surfaces. If the lubricating film's pressure is supplied by an external source, the system is said to be lubricated hydrostatically. If the pressure between the surfaces is generated as a result of the shape and motion of the surfaces themselves, however, the system is hydro dynamically lubricated. This second type of lubrication depends upon the viscous properties of the lubricant.

Boundary Lubrication. A condition that lies between unlubricated sliding and fluid–film lubrication is referred to as boundary lubrication, also defined as that condition of lubrication in which the friction between surfaces is determined by the properties of the surfaces and properties of the lubricant other than viscosity. Boundary lubrication encompasses a significant portion of the lubrication phenomena and commonly occurs during the starting and stopping of machines.

Solid Lubrication. Solids such as graphite and molybdenum disulfide are widely used when normal lubrication do not possess sufficient resistance to load or temperature extremes. But lubricants need not take only such familiar forms as fats, powders, and gases; even some metals commonly serve as sliding surfaces in some sophisticated machines.

Functions of Lubricants

Although a lubricant primarily controls friction and wear, it can and ordinarily does perform numerous other functions, which vary with the application and usually are interrelated.

Friction Control. The amount and character of the lubricant made available to sliding surfaces have a profound effect upon the friction that is encountered. For example, disregarding such related factors as heat and wear but considering friction alone between two oil-film lubricated surfaces, the friction can be 200 times less than that between the same surfaces with no lubricant. Under fluid-film conditions, friction is directly proportional to the viscosity of the fluid. Some lubricants, such as petroleum derivatives, are available in a great range of viscosities and thus can satisfy a broad spectrum of functional requirements. Under boundary lubrication conditions, the effect of viscosity on friction becomes less significant than the chemical nature of lubricant.

Wear Control. Wear occurs on lubricated surfaces by abrasion, corrosion, and solid-to-solid contact. Proper lubricants will help combat each type. They reduce abrasive and solid-to-solid contact wear by providing a film that increases the distance between the sliding surfaces, thereby lessening the damage by abrasive contaminants and surface asperities.

Temperature Control. Lubricants assist in controlling temperature by reducing friction and carrying off the heat that is generated. Effectiveness depends upon the amount of lubricant supplied, the ambient temperature, and the provision for external cooling. To a lesser extent, the type of lubricant also affects surface temperature.

Corrosion Control. The role of lubricants in controlling corrosion of the surfaces themselves is twofold. When machinery is idle, the lubricant acts as a preservative. When machinery is in use, the lubricant controls corrosion by coating lubricated parts with a protective film that may contain additives to neutralize corrosive materials. The ability of a lubricant to control corrosion is directly related to the thickness of the lubricant film remaining on the metal surfaces and the chemical composition of the lubricant.

Other Functions

Lubricants are frequently used for purposes other than the reduction of friction. Some of these applications are described below.

Power Transmission. Lubricants are widely employed as hydraulic fluids in fluid transmission devices.

Insulation. In specialized applications such as transformers and switchgear, lubricants with high dielectric constants act as electrical insulators. For maximum insulating properties, a lubricant must be kept free of contaminants and water.

Shock Dampening. Lubricants act as shock-dampening fluids in energy transferring devices such as shock absorbers and around machine parts such as gears that are subjected to high intermittent loads.

Sealing. Lubricating grease frequently performs the special function of forming a seal to retain lubricants or to exclude contaminants.

◇ New Words and Expressions

sliding ['slaɪdɪŋ]	n. 滑，移动；adj. 变化的，滑行的
lubricant ['luːbrɪk(ə)nt]	n. 润滑剂，润滑材料
unlubricated ['ʌn'l(j)uːbrɪkeɪtɪd]	adj. 无润滑的
seize [siːz]	v. 卡住，咬住，黏结
weld [weld]	n. 焊接，熔接
cleanliness ['klenlɪnəs]	n. 清洁度，洁净
adsorb [əd'sɔːb]	vt. 吸附，吸取
contaminant [kən'tæmɪnənt]	n. 污染物，致污物
intentionally [ɪn'tenʃənəli]	adv. 故意地
hydrostatical [haɪdrəu'stætɪkəl]	adj. 流体静力学的
hydrodynamical [haɪdrouda'næmɪkəl]	adj. 流体的，流动的
viscosity [vɪ'skɒsɪtɪ]	n. [物] 黏性，[物] 黏度
encompass [ɪn'kʌmpəs; en-]	vt. 包含，包围，环绕，完成
graphite ['græfaɪt]	n. 石墨
molybdenum [mə'lɪbdənəm]	n. [化学] 钼
disulfide [daɪ'sʌlfaɪd]	n. [无化] 二硫化物
temperature extreme	温度极限
fat [fæt]	n. 脂肪，油脂；adj. 油脂的，多脂的
profound [prə'faʊnd]	adj. 深厚的，意义深远的，渊博的
derivative [dɪ'rɪvətɪv]	n. 衍生物，派生物
spectrum ['spektrəm]	n. 光谱，频谱，范围，余象
abrasion [ə'breɪʒ(ə)n]	n. 磨损，磨耗，擦伤
corrosion [kə'rəʊʒ(ə)n]	n. 腐蚀，腐蚀产生的物质，衰败
lessen ['les(ə)n]	v. 减少，减轻，缩小
asperity [ə'sperɪtɪ]	n. (表面的) 粗糙，凹凸不平
ambient ['æmbɪənt]	n. 周围环境；adj. 周围的，外界的
provision [prə'vɪʒ(ə)n]	n. 措施，保证，保障
to a lesser extent	在较小的程度上
preservative [prɪ'zɜːvətɪv]	n. 防腐剂，预防法，防护层
additive ['ædɪtɪv]	n. 添加剂，添加物
neutralize ['njuːtrəlaɪz]	vt. 使……中和，使……中立
insulation [ɪnsjʊ'leɪʃ(ə)n]	n. 绝缘，隔离，孤立

transformer [træns'fɔːmə]   n. [电] 变压器
switchgear ['swɪtʃgɪə]   n. 开关设备，接电装置
dielectric [ˌdaɪɪ'lektrɪk]   adj. 不导电的，绝缘的
dielectric constant   介电常数，介质常数
dampen ['dæmp(ə)n]   vt. 抑制，使……沮丧，丧气
absorber [əb'sɔːbə]   n. 减振器，吸收器，吸收体
intermittent [ɪntə'mɪt(ə)nt]   adj. 间歇的，断断续续的，间歇性
sealing ['siːlɪŋ]   n. 密封件，密封，封闭
grease [griːs]   n. 润滑脂，黄油
retain [rɪ'teɪn]   vt. 保持，保留，保有
exclude [ɪk'skluːd]   vt. 排除，排斥，拒绝接纳，逐出

Questions

1. What types of lubrication are there?
2. What are functions of lubricants?
3. What is boundary lubrication?
4. Apart from reduction of friction, what other purposes do lubricants have?

Unit 23 Kawasaki Power Loader

Job-proven, Rugged Cummins Engine

This four-cycle, six-cylinder, water-cooled, turbocharged direct-injection engine has been designed for maximum performance and efficiency. It develops a consistently high torque across a wide range of engine rpm's while utilizing an efficient fuel-air mixture combustion, delivering 450 HP (gross) to meet all loader operating demands of its class[1]. A large fuel tank (620ltr) enables long operating hours without refilling.

Computer-controlled Automatic Transmission

No more annoying gear shifts! Once you select the direction and AUTO with the directional and speed control lever, the optimum engine rpm's and speed (forward: 2nd, 3rd and 4th/reverse: 2nd and 3rd) are automatically selected by the computer-control system (see Figure 23-1), leaving you to concentrate on bucket loading operations only.

Figure 23-1 Computer-controlled automatic transmission

Dependable Service Brake

Braking is assured with dual-line hydraulically-actuated, fully-sealed wet multi-disc brakes, so highly responsive dependable braking is delivered to all four wheels with light-touch operation[2]. And a wear-resistant lining is used for the discs to prolong the service life. (see Figure 23-2)

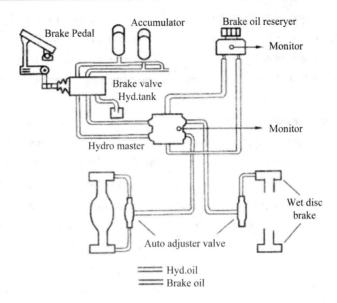

Figure 23-2　Dependable service brake

Tough Frame, Boom and Bucket Cylinder Bracket

Full box section main frame has been strengthened to support engine and power train component, while the boom and bucket cylinder brackets are made of quality thick plates to resist shock and loading stresses.

Simplified Checks and Maintenance

One-touch swing-up radiator grille—To facilitate cleaning the radiator, the radiator grille is fully-opened at the top, particularly useful on sites where clogging occurs and immediate cleaning is needed.

Wide-opening engine compartment—The engine side covers can be opened wider, making it easy to check the engine and related parts from the ground.

Simplified engine oil drainage—Engine oil drain plug is located outside the engine compartment for easy access without getting dirty.

Double-structure air cleaner—The double element air cleaner filters the contaminated outside air to supply pure air to the engine. Accumulated dust is automatically removed when the engine stops.

Sealed loader linkage pins—All pins are fully sealed with grease to provide dependable service with minimum maintenance. Metal-seal rings also extend the life of the pins and bushings.

A Comfortable Operator Housed in A Rugged Machine Performs at His Peak

Multi–adjustable operator's seat—The 115ZIV provides a wealth of functional comforts. The bucket seat with armrests is fully adjustable for height, position and tilt, and the critical balance between seat location, visual position, steering wheel, pedals and levers is carefully designed. The padding inside the fabric seat cover is designed to absorb extra vibration.

Lower noise—Noise inside the operator's compartment is minimized by the extra margin of power in the CUMMINS engine, a large exhaust muffler, soundproofing urethane rubber lining inside the engine, and vibration-absorbing rubber.

Amenities and arm-rest—All the above ensure lasting comfort during long hours at the wheel. What's more, many extra features such as a drink holder, trays, storage box, etc. are arranged within easy reach of the operator. A comfortable sliding and vertically-adjusted arm-rest on the right side of the operator's seat reduces fatigue during busy boom and bucket operations.

Easy-access operator controls—Naturally the directional and gearshift lever, accelerator pedal, brake pedals, bucket and boom control levers, handle-mounted steering wheel and selector switches and buttons are ergonomically designed and easy to reach [3].

Single-lever directional and gearshift control—With a single lever on the left of the steering column, all directional and gearshift changes are brought to you instantly; just select the direction by pulling the lever toward you for reverse or push it away for forwards and turn the lever to the desired forward or reverse speed [4].

Forward: 1st, 2nd, 3rd and AUTO (2nd, 3rd and 4th automatically changed)

Reverse: 1st, 2nd and 3rd (2nd and 3rd automatically changed)

Panoramic cab (option)—A new panoramic cab (option) not only protects the operator from external hazards but also provides an excellent working environment. To ensure excellent vision, the front and rear windshield are curved glass and intermittent windshield wipers are provided. The optimum mix of cool and warm air is ensured with a new air conditioning system (option). The cab interior is another feature; the felt-lined ceiling, thick rubber-matted floor and various amenities are coordinated in a tasteful color.

Light-touch steering and assured loading—The orbitrol load-sensing system supplies hydraulic pressure to the steering (takes priority over loading system) only if needed so that light-touch steering is ensured even at low engine speeds. When steering is not required, the full discharge from the steer oil pump is applied to the loading circuit so that powerful loading is ensured for greater energy savings.

Assured precision braking—In addition to the brake-only pedal (right), the declutch pedal (left) is used for precise braking while the transmission is in neutral by turning on the transmission cut-off switch, thus high rpm's can be maintained for bucket loading operations. Of course, this pedal can be used for braking only when the transmission cut-off switch is turned off, to ensure safe operation on slopes or ramps.

At-a-glance meters and gauges—The exact conditions of the machine can be instantly checked from the meters, gauges and indicator lights on the instrument panel; speedometer, engine water temp, gauge, fuel indicator, air pressure gauge, engine hour meter, and various other warning and indicator lights as well as selector switches and buttons.

PUS(power-up switch)—The boom control lever is accompanied by a small PUS lever to accommodate gearshifts from 2nd to 1st and vice versa for scooping material easily without taking your hand off the bucket control lever.

Optimum bucket control detent—The bucket control lever allows a full bucket to be scooped very easily because a detent is activated at the position where the bucket encounters extra load while raising the boom so the bucket and boom can move smoothly and efficiently.

◆ *New Words and Expressions*

proven ['pru:vn]	adj. 被证实的
rugged ['rʌgɪd]	adj. 坚固的
bucket ['bʌkɪt]	n. 铲斗
lining ['laɪnɪŋ]	n. 内衬
grille [grɪl]	n. 百叶窗，护栅
drainage ['dreɪnɪdʒ]	n. 排放
bushing ['bʊʃɪŋ]	n. 衬套
ergonomically [ɜːgəʊ'nɒmɪkəli]	adv. 人类工程学地
amenity [ə'miːnəti]	n. 舒适，适宜
fatigue [fə'tiːg]	n. 疲劳
panoramic [ˌpænə'ræmɪk]	adj. 全景的
hazard ['hæzəd]	n. 伤害，损害
detent [dɪ'tent]	n. 棘爪
gear shift	换挡
linkage pin	连接销
light-touch	轻触式，触摸式
across a wide range of	在……各种范围内
meet all demands	满足各种要求
a wealth of	大量的，许多的
take priority over/of	优先于……

⇨ *Notes*

[1]　It develops a consistently high torque across a wide range pf engine rpm's while utilizing an efficient fuel-air mixture combustion, delivering 450 HP (gross) to meet all loader operating demands of its class.

参考译文：在各种发动机转速条件下，始终都有高转矩，同时通过高效地燃-气混合气的燃烧，能产生 450 马力(有效的)，以满足同类装载机的各种施工要求。

delivering 引导的分词短语并不与 utilizing 分词短语并列，而是修饰该短语。

[2]　Braking is assured with dual-line hydraulically-actuated, fully-sealed wet multi-disc brakes, so highly responsive dependable braking is delivered to all four wheels with light-touch

operation.

参考译文：双回路液压启动的全封式浸油多盘的车闸保证装载机制动万无一失，因此轻轻按一下按钮就能制动四个灵敏可靠的车轮。

hydraulically-actuated 和 fully-sealed 是英语科技文章中常见的复合词形式，因为这种形式比定语从句更简洁，但被动关系比较难理解。将它展开成定语从句的形式，就比较好理解了，如本例句可展开为：…dual line wet multi-disc brakes that are actuated hydraulically and sealed fully…。

[3] Naturally the directional and gearshift lever, accelerator pedal, brake pedals, bucket and boom control levers, handle-mounted steering wheel and selector switches and buttons are ergonomically designed and easy to reach.

参考译文：当然，方向盘和换挡控制杆、油门踏板、刹车脚踏、铲斗和悬臂控制杆、手柄式方向盘以及选择开关和按钮，都是按照人类工程学原则设计的，操作起来得心应手。

像 ergonomically 这样表示学科的副词，往往不能直接翻译成汉语的副词，如"人类工程学地"，而应根据上下文将这种副词译成短语，如"按照人类工程学的原则"。

[4] Just select the direction by pulling the lever toward you for reverse or push it away for forwards and turn the lever to the desired forward or reverse speed.

参考译文：只需拉动控制杆就可选择方向：向里拉就后退，向外推就前进，同时转动控制杆就能选择所需的前进或后退的速度。

forwards 是副词，但在本例句中却作名词用，当介词 for 的宾语。英语科技文章，尤其是在说明书中有时为了文字简洁明了，可能会打破一般的语法规则。

☒ Questions

1. What is the function of dependable service brake?
2. What special features does Kawasaki Power Loader have?
3. What does PUS stand for?
4. What is the purpose of using the declutch pedal?
5. Why is the radiator grille fully-opened at the top?

Reading Material: Computer Technology

Introduction

The central and essential ingredient of CAD/CAM is the digital computer[1]. Its inherent speed and storage capacity have made it possible to achieve the advances in image processing, real-time process control, and a multitude of other important functions that are simply too complex and time consuming to perform manually. To understand CAD/CAM, it is important to be familiar with the concepts and technology of the digital computer.

The modern digital computer is an electronic machine that can perform mathematical and logical calculations and data processing functions in accordance with a predetermined program of instructions. The computer itself is referred to as hardware, whereas the various programs are referred to as software.

There are three basic hardware components of a general-purpose digital computer:
(1) Central processing unit (CPU)
(2) Memory
(3) Input/output (I/O) section

The relationship of these three components is illustrated in Figure 23-3. The central processing unit is often considered to consist of two subsections: a control unit and an arithmetic-logic unit (ALU). The control unit coordinates the operations of all the other components. It controls the input and output of information between the computer and the outside world through the I/O section, synchronizes the transfer of signals between the various sections of the computer, and commands the other sections in the performance of their functions. The arithmetic-logic unit carries out the arithmetic and logic manipulations of data. It adds, subtracts, multiplies, divides, and compares numbers according to programmed instructions. The memory of the computer is the storage unit. The data stored in this section is arranged in the storage unit of the computer for processing. Finally, the input/output provides the means for the computer to communicate with the external world. This communication is accomplished through peripheral equipment such as readers, printers, and process interface devices. The computer may also be connected to external storage units (e.g., tapes, disks, etc.) through the I/O section of the computer.

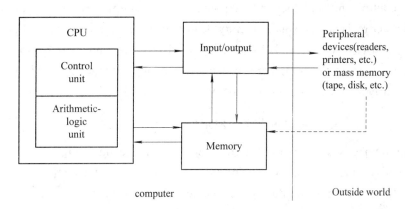

Figure 23-3   Basic hardware structure of a digital computer

The software consists of the programs and instructions stored in memory and in external storage units. It is the software that assigns to the computer the various functions which the user desires the system to accomplish. The usefulness of the computer lies in its ability to execute the changed, and therefore different programs can be placed into memory, the digital computer can be used for a wide variety of applications.

Regardless of the application, the computer executes the program through its ability to manipulate data and numbers in their most elementary form. The data and numbers are represented which is called the binary system. The more familiar decimal number system and a whole host of software languages can utilize the binary system to permit communication between computers and human beings.

Central Processing Unit (CPU)

The central processing unit (CPU) regulates the operation of all system components and performs the arithmetic and logical operations on the data. To accomplish these functions, the CPU consists of two operating units:

(1) Control unit
(2) Arithmetic-logic unit (ALU)

The control unit coordinates the various operations specified by the program instructions. These operations include receiving data which enter the computer and deciding how and when the data should be processed. The control unit directs the operation of the arithmetic-logic unit. It sends data to the ALU and tells the ALU what functions to perform on the data and where to store the results. The capability of the control unit to accomplish these operations is provided by a set of instructions called an executive program which is stored in memory.

The arithmetic and logic unit performs operations such as addition, subtractions, and comparisons. These operations are carried out on data in binary form. The logic section can also be used to alter the sequence in which instructions are executed when certain conditions are indicated and to perform other functions, such as editing and masking data for arithmetic operations.

Both the control unit and the arithmetic-logic unit perform their functions by utilizing registers. Computer registers are small memory devices that can receive, hold, and transfer data. Each register consists of binary cells to hold bits of data. The number of bits in the register establishes the word length the computer is capable of handling. The number of bits per word can be as few as 4 (early microcomputers) or as many as 64 (large scientific computers).

Computer Programming Languages

The binary number system could be used to represent any decimal number, alphabetic letter, or other common symbol. Data and instructions are communicated to the computer in the form of binary words. In executing a program, the computer interprets the configuration of bits as an instruction to perform electronic operations such as add, subtract, load into memory, and so forth. The sequence of these binary-coded instructions defines the set of calculations and data manipulations by which the computer executes the program.

The binary-coded instructions that computers can understand are called machine language. Unfortunately, binary-coded instructions and data are very difficult for human programmers to read or write. Also, different machines use different machine languages which can be learned with relative ease by human beings. In all, there are three levels of computer programming languages:

(1) Machine language
(2) Assembly language

(3) Procedure-oriented (high-level) language

Machine and Assembly Languages

The language used by the computer is called machine language. It is written in binary, with each instruction containing an operation code and an operand. The operand might be a memory address, a device address, or data. In machine language programming, storage locations are designated for the program and data, and these are used throughout the program to refer to specific data or program steps. In addition, the programmer must be familiar with the specific computer system since machine language instructions are different for each computer. Programming in machine language is tedious, complicated, and time consuming. To alleviate the difficulties in writing programs in binary, symbolic languages have been developed which substitute English-like mnemonics for each binary instructions. Mnemonics are easier to remember than binary, so they help speed up the programming process. A language consisting of mnemonic instructions is called an assembly language.

Assembly languages are considered to be low-level languages. The programmer must be very knowledgeable about the computer and equipment being programmed. Low-level languages are the most efficient in terms of fast execution on the computer, but there are obvious difficulties for the programmer in writing large programs for various applications using different computers.

Assembly language programs must be converted into machine language before the computers can execute them. The conversion is carried out by a program called an assembler. The assembler takes the assembly language program, performs the necessary conversions, and produces two new programs: the machine language version and an assembly listing. The assembly listing shows the mnemonic instructions and their associated machine language equivalents, and any errors the original assembly language program may have contained.

High-level Languages

Assembly language is machine oriented. High-level languages, by contrast, are procedure oriented. They are to a large extent independent of the computer on which they are used. This means that a program written on one computer can be run on a different computer without significant modifications to the program.

High-level languages consist of English-like statements and traditional mathematical symbols. Each high-level statement is equivalent to many instructions in machine language.

The advantage of high-level languages is that it is not necessary for the programmer to be familiar with machine language. The program is written as an English-like algorithm to solve a problem. Like assembler languages, high-level languages must also be converted into machine code. This is accomplished by a special program called a compiler. The compiler takes the high-level program, and converts it into a lower-level code, such as the machine language. If there are any statement errors in the program (e.g., misspelled words), error messages are printed in a special program listing by the compiler.

There are many different high-level languages. Some of the most commonly used of these are VC, VB, MATALB and so on.

Computer Process Interfacing

To be useful, the computer must be capable of communicating with its environment. In a data processing system, this communication is accomplished by the various input/output devices such as card readers, printers, and CRT consoles. Functioning as a process control system, the computer must be capable of sensing the important process variables from the operation and providing the necessary responses to maintain effective control over the process. In the following sections we examine some of the important components of a computer process control system.

(1) Manufacturing Process Data: the data that must be communicated between the manufacturing process and the computer. These data can be classified into three categories:
- Continuous analog signals
- Discrete binary data
- Pulse data

Continuous analog data can be represented by a variable that assumes a continuum of values over time. During the manufacturing process cycle, it remains uninterrupted, and the values it can assume are restricted to a finite range. Examples of analog variables include temperature, pressure, liquid flow rate, and velocity. Each of these phenomena are continuous functions over time and are capable of taking on an infinite number of values in a certain range. The number of values permitted is a function of the ability of measuring instruments to distinguish between different signal levels.

Discrete binary data can take on either of two possible values, such as on and off, or open and closed. Switches, motors, valves, and lights are all devices whose status at any time may be a binary function. Binary data are represented in electronic digital systems as typically one of two voltage levels whose values depend on the specific devices that make up the system. Typical voltage levels are 0 and +5V.

Pulse data consist of a train of pulsed electrical signals from devices called pulse generators. The pulse train can be used to drive devices such as stepping motors. The magnitude of each pulse is fixed; the magnitude of the pulse train is the number of pulses it contains. Since the number of pulses in a serial can be counted over a period of time, that number can be represented as digital data. Conversely, digital data can be used to produce a pulse train of a given magnitude.

(2) System Interpretation of Process Data: the three categories of manufacturing process data must be capable of interacting with the computer. For monitoring the process, input data must be entered into the computer. For controlling the process, output data must be generated by the computer and converted into signals understandable by the manufacturing process. There are six categories of computer-process interface representing the inputs and outputs for the three types of process data. These categories are: Analog to digital, Contact input, Pulse counters, Digital to analog, Contact output, Pulse generators

Analog-to-digital interfacing involves transforming real-valued signals into digital representations of their magnitude. A number of different steps must be accomplished to effect this conversion process. These steps involve the following hardware:

a. Transducers, which convert a measurable process characteristic (flow rate, temperature, process, etc.) into electrical voltage levels corresponding in magnitude to the state of the characteristic of the process measured. A thermocouple is an example of a device in this category. It converts temperature into a small voltage level.

b. Signal conditioners, which filter random electrical noise and smooth the analog signal emanating from transducing devices [2].

c. Multiplexers, which connect several process-monitoring devices to the analog-to-digital converter. Each process signal is sampled at periodic intervals and passed on to the converter.

d. Amplifiers scale the incoming signal up or down to the range of the incoming real-valued process signals into their digital equivalents.

e. The analog-to-digital converter (ADC) transforms the incoming real-valued process signals into their digital equivalents.

f. The digital computer's I/O section accepts the digital signals from the ADC. A limit comparator is often connected between the I/O section and the ADC to prevent out-of-limit signals from distracting the CPU.

The contact input interface is a set of simple contacts that can be opened and closed to indicate the status of limit switches, button positions, and other binary-type data. It serves as the intermediary between discrete process data and the computer, which periodically scans the signal status and compares it with preprogrammed values.

Digital transducers belong to a class of electronic measuring instruments that generate as output a series of electrical pulses of uniform magnitude. A pulse counter is used to convert the pulse trains into a digital representation, which is then applied to the computer's input channel[3]. The last three devices discussed are concerned with processing data from the computer to the process.

The digital-to-analog converter takes digital data generated by the computer and transforms it into a pseudo analog signal. The signal is considered pseudo analog because the computer is only capable of processing a limited-precision digital word, so that an infinite number of analog signal levels cannot be generated.

Contact output interfaces, like their input counterparts, are sets of contact that can be opened or closed. The output word of the computer is used to turn to indicator lights, alarms, and even equipment functions such as cutting oil pumps.

Computer Network Structures

The term "computer network" refers to the actual physical connections between computers in the structure. This structure consists not only of the computers, but also the terminals, transmission lines, switching centers, and other devices required to accomplish the desired functions of the system. The number of possible configurations is limitless. However, the various network arrangements divide themselves into several basic categories, sometimes called topologies; four of the more common topologies are illustrated in Figure 23-4.

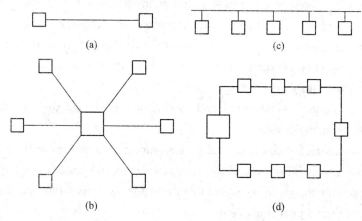

Figure 23-4 Types of computer network structures

(a) point-to-point   (b) star   (c) multidrop   (d) loop

In the point-to-point configuration, a communication line is established between two devices (typically, a computer and devices is limited). However, when there are a large number of processors in the structure, the number of point-to-point communication links can become significantly large, resulting in a relatively expensive system. Other network structures share the communication links in various ways to avoid this expense. An inherent advantage of the point-to-point configuration is that whenever one of the direct links is broken, the other connections remain intact.

The star network consists of a central computer and several peripheral devices connected to it. The central computer is referred to as the master and the peripheral devices connected to function of the master is to delegate tasks to each of the smaller units or act as a switch to connect various units. One of the important deficiencies of the star configuration is that the entire network system is put out of operation if the central computer fails.

In the multi-drop network structure, computers and other devices are connected by a single communication line. This line is either a bus or a serial channel. The limitation of this arrangement is that only two processors are allowed to transmit data to each other at the same time. This results from the shared nature of the communication lines [4].

In a loop configuration, each computer is connected to a communication line which begins and terminates at a loop controller, typically a computer that maintains control over inter computer communications. Messages to a specific computer are usually transmitted to all computers in the network, with the addressee decoding its address and accepting the information[5].

◇ *New Words and Expressions*

ingredient [ɪnˈgriːdiənt]   n. 组成部分，原料

multitude ['mʌltɪtuːd]	n. 大量，许多
predetermine [ˌpriːdɪ'tɜːrmɪn]	vt. & vi. 预先裁定，注定
refer to	涉及
digital computer	数字计算机
in accordance with	与……一致，依照

⇨ Notes

[1] The central and essential ingredient of CAD/CAM is the digital computer.

参考译文：CAD/CAM 是指计算机辅助设计与计算机辅助制造集成系统；CAD/CAM 的核心部分是计算机。

[2] Signal conditioners, which filter random electrical noise and smooth the analog signal emanating from transducing devices.

参考译文：信号调整器，它可滤掉杂乱的电噪声，并可修整由传感装置送出的模拟信号。

[3] A pulse counter is used to convert the pulse trains into a digital representation, which is then applied to the computer's input channel.

参考译文：脉冲计数器可将脉冲串转换为一种数字代码，以适合于计算机输入通道。

be applied to 意为"适用于"。

[4] This results from the shared nature of the communication lines.

参考译文：这来自通信线的共享特性。

[5] Messages to a specific computer are usually transmitted to all computer in the network, with the addressee decoding its address and a accepting the information .

参考译文：送往某特定计算机的信息，一般都传送到网络中所有计算机，同时由信息接收计算机译出地址码并接收信息。

with 意为"同时"。

⇨ Questions

1. What is the definition of the modern digital computer?
2. What are the three basic hardware components of a general-purpose digital computer?
3. Can you give some examples of different high-level computer languages?
4. What is the function of CPU? What unit does it consist of?
5. What does the term "computer network" refer to?

Unit 24 Komatsu Truck

Excellent Productivity—Power and Speed on the Steepest Inclines

High-output Komatsu SA6D140 engine: The 15.2-liter power plant with turbocharger and aftercooler develops a massive output of 488 HP(364 kW) at 2100 RPM [1].

7-speed, fully automatic K-ATOMiCS transmission: The K-ATOMiCS (Komatsu Advanced Transmission with Optimum Modulation Control System) automatically selects the optimum gear according to vehicle speed, engine RPM and the shift position you've chosen [2]. The result: the best gear for any driving situation.

Oil-cooled multiple-disc retarder and optional exhaust retarder: The truck can be decelerated without frequent use of the brakes, allowing you to travel safer at higher speeds, even down long, steep slopes [3].

A Move Stable Ride in A More Maneuverable Truck

Long wheelbase and wide tread: With an extra-long wheel-base, a wide thread and an exceptionally low center of gravity, the HD325-6 hauls the load at higher speed for more production, and delivers supreme driving comfort over rough terrain.

Big body: A wide target area of 19 m^2 (204 sq.ft) makes for easy loading with minimal soil spillage and more efficient hauling.

Small turning radius: The MacPherson strut type front suspension system has a special frame between each wheel and the main frame. The wider space created between the front wheels and the main frame increases the turning angle of the wheels. The larger this turning angle, the smaller the turning radius of the truck.

Excellent Fuel Economy

Low fuel-consuming engine: High injection pressure creates an ideal fuel-air mixture for more combustion efficiency, while the ductile cast-iron pistons greatly reduce friction loss. For even more combustion efficiency, each cylinder has four valves—two for intake, two for exhaust. The two intake ports (both are helical type) produce optimum swirl for excellent combustion.

The exhaust gas is smoothly and quickly ejected from the combustion chamber through the exhaust ports. All this helps to make the Komatsu-built engine a fuel miser.

Enhanced Operation Comfort

K-ATOMiCS—smooth acceleration/deceleration: An electronically controlled valve is provided for each clutch pack in the transmission, allowing independent clutch engagement/disengagement. Moreover, it enables an ideal change in clutch modulation pressure and torque cut-off timing in response to traveling conditions. The result is smooth shifting and responsive acceleration.

Hydro pneumatic suspension: All four wheels have hydro pneumatic suspension with a fixed throttle damper control valve that greatly reduces pitching, rolling and bouncing over rough terrain.

Three-mode hydro pneumatic suspension (optional): To further enhance the driving comfort, automatic three-mode suspension is optionally available. It enables the operator to select one of three cushioning effects (SOFT, NORMAL or HARD) depending on the road conditions, for improved damping control [4].

Study, refined frame: Cast-steel components are employed in the main frame in critical areas where loads and shocks are most concentrated.

Adjustment-free brakes: The front service brakes and the parking brake are adjustment-free caliper disc type [5].

Easy maintenance: Greasing points have been centralized at three locations. Fuel and engine oil filters are also located together on the left-hand remote mount, for easy, remote inspection from the ground.

Rigorous dump body design: The dump body is made of 130kg/mm^2 (184,900 PSI) high-tensile-strength steel for excellent rigidity and reduced maintenance costs. The V-shape design also increases structural strength. The side and bottom plates of the dump section are reinforced with ribs for added strength.

Monitor: A monitor system informs the operator of any abnormality at two levels: CAUTION and EMERGENCY [6]. Since failures are detected before they become critical, the HD3256 is more reliable and safer than ever.

Payload meter: In addition to a production-amount display, the payload meter indicates loading status to the loader operator with caution lamps on the side of the cab. Not only is overloading prevented, but the machine can be more closely controlled for more productivity and economy.

Rock body: The optional rock body, with reinforcement liners fitted throughout the body interior, is available for heavy-duty rock loading assignments at quarry, limestone mine and other tough jobsite.

Engine exhaust retarder: It improves safety and hauling performance.

Three-mode hydro pneumatic suspension (auto-suspension): Damping force is automatically switched to one of three stages (SOFT, NORMAL and HARD) according to load and road conditions, for a more comfortable and stable ride.

Monitor with maintenance control function: The monitor automatically checks items to be checked before starting, so daily maintenance is carried out easily. Filter and oil replacement times are also suggested for more foolproof maintenance.

ROPS: It protects the operator and the cab from damage should the truck turn over.

◇ New Words and Expressions

turbocharger ['tɜ:bəʊtʃɑ:dʒə(r)]	n. 涡轮增压 (机构)
optimum ['ɒptɪməm]	adj. 最佳的
modulation [ˌmɒdjʊ'leɪʃn]	n. 调制，调控
massive ['mæsɪv]	adj. 强大的
retarder [rɪ'tɑ:də]	n. 减速器
wheelbase ['wi:lbeɪs]	n. 车轮轴距
injection [ɪn'dʒekʃn]	n. 注射，喷射
cushioning ['kʊʃənɪŋ]	n. 减震，缓冲
abnormality [ˌæbnɔ:'mæləti]	n. 异常
turning radius	转弯半径
parking brake	手闸
make for	有利于
in response to	响应，根据
be centralized at	集中在
be switched to	转换到

⇨ Notes

[1] The 15.2-liter power plant with turbocharger and aftercooler develops a massive output of 488 HP (364 kW) at 2100 RPM.

参考译文：15.2L 功率的发动机，配备有涡轮增压和后冷却系统，每分钟 2100 转，488 马力 (364 kW) 强大输出功率。

plant 意为"设备"，指代前面的 engine 一词。with 可理解为 equipped with，意为"配备有"，在科技文献中这是一种常见的修饰手段。

[2] The K-ATOMiCS (Komatsu Advanced Transmission with Optimum Modulation Control System) automatically selects the optimum gear according to vehicle speed, engine RPM and the shift position you've chosen.

参考译文：K-ATOMiCS(配有最佳调控系统的 Komatsu 传动系统)可根据您选择的车速、发动机转速和挡位，自动选择最佳传动比。

K-ATOMiCS 是 Komatsu Advanced Transmission with Optimum Modulation Control

System 的首字母缩写单词，可按一般的单词读音，/ka-təu-miks/，正因如此，在此单词中添加了一个字母"i"，以便发音。另如：NATO /neitəu/ (North Atlantic Treaty Organization，北大西洋公约组织)，UNESCO /ju:-nes-əu/ (United Nations Educational, Scientific and Cultural Organization，联合国教科文组织)。对于广泛接受的并为人熟悉的缩写词，可采用现在的翻译或直接用缩写。如果是文章中特有的缩写，可译成中文的缩写或直接用英文的缩写，但第一次一定要注明其英文的全称。

[3] The truck can be decelerated without frequent use of the brakes, allowing you to travel safer at higher speeds, even down long steep slopes.

参考译文：本卡车不需频繁使用车闸就能减速，使得您即使在下长长的陡坡时也能更安全地、更高速地行驶。

科技文献中常用 allowing 和 enabling 引导的分词短语修饰前面整句话的内容，表示结果，意为"从而使……变成可能"。又如：An electronically controlled valve is provided for each clutch pack in the transmission, allowing independent engagement/disengagement. (传动系统的离合器组件配有电子控制阀，使得各离合器能独立啮合和断开)。

[4] It enables the operator to select one of three cushioning effects (SOFT, NORMAL or HARD) depending on the road conditions, for improved damping control.

参考译文：使操作员能够根据路况从三种减震方式 (柔软、正常、坚硬) 中选择其一，以达到更好地控制减震的效果。

depending on 引导的分词短语常用来表示不同的选择，不同的变化的条件依据，意为"根据……"，类似于 according to 和 in response to。另如：The engine can operate at different speeds depending on the work tasks. (发动机能根据施工任务的不同以不同的速度运行)。

[5] The front service brakes and the parking brake are adjustment-free caliper disc type.

参考译文：前端脚闸和手闸都属免调卡盘式制动装置。

adjustment-free 通过连字符与名词连接构成复合词，做形容词用，其中 free 表示"免除"和"无"的意义，如：duty-free (免税的)，friction-free (无摩擦的)。有类似用法的形容词还有 proof, resistant, friendly, 如：foolproof (防错的)，waterproof (防水的)，corrosion-resistant (防腐的)，user-friendly (用户界面友好的)。本文中还出现了形容词加名词复合词，heavy-duty (重负荷的)。在前一类复合词中，之所以将形容词置于名词后，是因为名词与形容词的关系原本需要介词连接，如：free of duty, free of friction, proof against water, resistant against corrosion, friendly to user。如果将形容词放在名词前，则意义要么不通，要么就不一样，如：free duty (免费税)，friendly user (友好的使用者)。

[6] A monitor system informs the operator of any abnormality at two levels: CAUTION and EMERGENCY.

参考译文：监视系统给操作员提供的卡车异常的信息分为两个级别："警告"和"紧急"。英文使用说明中的大写单词，常常表示设备上有此文字标识，汉语译文中一般加引号表示。

⊠ Questions

1. What is low fuel-consuming engine?

2. What is the function of payload meter?
3. What is the function of ROPS?
4. What is the function of three-mode hydro pneumatic suspension?

Reading Material: Automobile Components

Engine

The engine is a power plant, which provides power to drive the automobile.

In most automobile engines, the explosive power of the mixture of air and gasoline drives the pistons. The pistons turn a crankshaft to which they are attached. The rotating force of the crankshaft makes the automobile's wheels turn.

Some automobiles are powered by another kind of engine, known as the rotary valve, rotating combustion engine or Wankel engine. The rotary valve engine also draws in a mixture of air and fuel, which is then compressed and burnt. A motor revolving in an elliptical chamber is connected to a shaft, which finally drives the rear wheels. In most automobiles, the engine is mounted at the front end of the car, with the clutch and gearbox immediately behind it; the engine, clutch and gearbox are assembled into a single unit.

A number of systems are necessary to make an engine work. A lubrication system is needed to reduce friction and prevent engine wear. A cooling system is required to keep the engine's temperature within safe limits. The engine must be provided with the correct amount of air and fuel by fuel system.

The mixture of air and fuel must be ignited inside the cylinder at just the right time by an ignition system. Finally, an electrical system is required to operate the cranking motor that starts the engine and to provide electrical energy to power engine accessories [1].

Lubrication System

An engine has many moving parts which eventually develop wear, as they move against each other. The engine circulates oil between these moving parts to prevent the metal-to-metal contact that results in wear. Parts that are oiled can move more easily with less friction and hence power loss due to friction is minimized. The secondary function of lubricant is to act as a coolant and also as a sealing medium to prevent leakages. Finally, a film of lubricant on the cylinder walls helps the rings in sealing and thus improves the engine's compressions.

Cooling System

Due to the combustion of fuel with air inside the cylinder, the temperature of the engine parts increases. This increase of temperature directly affects the engine performance and the life of the engine parts. The cooling system keeps the engine operating at the efficient temperature. Whatever the driving conditions, the system is designed to prevent both overheating and overcooling.

Fuel System

The main function of the fuel supply system is to provide fuel to carburetor or injection system at a rate and pressure sufficient to meet engine demands under all conditions of load, speed and gradients encountered by the vehicle. The fuel system must also have enough reserve fuel for several miles of vehicle operation.

Ignition System

The purpose of the ignition system is to provide assistance for the combustion of fuel either by a high voltage spark or self-ignition in each of the engine of engine's cylinder at the right time so that the air-fuel mixture can burn completely.

The fuel supplied to the combustion chamber must be ignited to deliver power. In a spark-ignition engine an electric spark is used for this purpose. The compression-ignition engine does not engine an electric spark is used for this purpose. The compression-ignition engine does not require a separate ignition system because the ignition is affected by compression of the mixture to a high pressure.

Electrical System

The engine's electrical system provides energy to operate a starting motor and to power all the accessories. The main components of the electrical system are a battery, an alternator, a starting motor, ignition coil and heater.

Frame

The frame provides a foundation for engine and the body of the vehicle. The frame is constructed from square or box-shaped steel members strong enough to support the weight of the body and other components.

The automobile frame is usually made up of a number of member welded or riveted together to give the final shape. The engine is mounted on the frame rubber pads which absorb vibrations and also provide damping of these vibrations. Absorption and damping of vibrations protects passengers from discomfort caused by shocks.

The frame is supported on wheel axles by means of springs. This whole assembly is called the chassis.

Suspension System

The function of the suspension system is to absorb vibrations due to the up and down motion of wheels, caused by the irregularities in the road surface. The springs, connecting linkages, and shock absorber comprise the suspension system of a vehicle. The suspension system is of two system is of types:

(1) Rigid system

(2) Independent system

In the rigid system, the road springs are attached to a rigid beam axle. It is mostly used in the front axle of commercial vehicles and in the car axle of all types of vehicles.

The independent system does not have a rigid axle. Each wheel is free to move vertically without any reaction on its mating wheel. The independent system is mostly used in small cars.

Power Train

The power train carries the power that the engine produces to the car wheels. It consists of the clutch (on cars with a manual transmission), transmission (a system of gears that increases the turning effort of the engine to move the automobile), drive shaft, differential and rear axle.

Clutch

A clutch is required with the manual transmission system to temporarily disconnect the engine from wheels. Such disengagement of the power train from the engine is essential while changing the gear ratio or while stopping the vehicle[2].

Transmission

The main function of the transmission is to provide the necessary variation to the torque applied by the engine to the wheels. This is achieved by changing the gearing ratio between the engine output shaft and the drive shaft.

Drive Shaft

The drive shaft or propeller shaft connects the gearbox and the differential unit. The drive shaft has universal joints at its ends.

Differential

The function of the differential is to split the power received from the propeller shaft to the rear axle shaft. It allows the rear wheels to be driven at different speeds when the vehicle takes a bend or falls into a ditch.

Axles are the shafts on which road wheels are mounted. The road wheels are provided with the required drive through these axles.

Wheels

The automobile wheels take the load of the vehicle and also produce tractive force to move the vehicle. The wheels are also used for retardation and for stopping the vehicle.

Steering System

The steering system is used for changing the direction of the vehicle. The major requirements in any steering mechanism are that it should be precise and easy to handle, and that the front wheels should have a tendency to return to the straight-ahead position after a turn[3]. A gear mechanism, which is known as steering gear, is used in this system to increase the steering effort provided by the driver. This system makes the vehicle steering very easy as the driver does not have to put in much effort. Vehicle steering is not only required on a curved road but also while maneuvering on the busy traffic roads. The steering system allows the vehicle to be guided i.e. to be turned left or right.

Braking System

Brakes are required for slowing down or stopping a moving vehicle. The braking system may be operated mechanically or hydraulically. 95 percent of the braking systems in use today are of the hydraulic type.

All brakes consist of two members, one rotating and the other stationary. There are various means by which the two members can be brought in contact, thus reducing the speed of the vehicle.

The major components of the braking system are: brake pedal, master cylinder, wheel cylinder, brake drum, brake pipe, brake shoes, brake packing plant and linkages. As the load on the vehicle and the vehicle speed has increased according to recent trends, in modern days, the importance of the braking system has also increased and power brakes are now being preferred[4]. Power brakes utilize vacuum and air pressure to provide most of the brake—applying effort.

◇ *New Words and Expressions*

explosive [ɪkˈspləʊsɪv]	adj. 爆炸 (性) 的；n. 爆炸物，炸药
wankel engine	n. 转子发动机
elliptical [ɪˈlɪptɪk(ə)l]	adj. 椭圆的，椭圆形的
gearbox [ˈgɪəbɒks]	n. 齿轮箱，变速箱
accessory [əkˈses(ə)rɪ]	n. 附件
circulate [ˈsɜːkjʊleɪt]	v. 使流通，使运行，使循环
coolant [ˈkuːl(ə)nt]	n. 冷却剂，冷冻剂
leakage [ˈliːkɪdʒ]	n. 漏，泄露，渗漏
compression [kəmˈpreʃ(ə)n]	n. 压缩，简约，压力
carburetor [ˌkɑːbjʊˈretə]	n. 化油器，汽化器
sufficient [səˈfɪʃ(ə)nt]	adj. 充分的，足够的
gradient [ˈgreɪdɪənt]	adj. 倾斜的；n. 梯度，倾斜度，坡
combustion chamber	燃烧室
spark-ignition	火花点火
compression-ignition	压燃

component [kəm'pəʊnənt]	n. 成分；adj. 组成的，构成的
alternator ['ɔːltəneɪtə]	n. 交流发电机，交替符
damping [dæmpɪŋ]	n. 阻尼，减幅，衰减
chassis ['ʃæsɪ]	n. 底盘
suspension [sə'spenʃ(ə)n]	n. 悬挂，吊，悬架
irregularity [ɪˌregjʊ'lærɪti]	n. 不规则，无规律
rigid ['rɪdʒɪd]	adj. 刚硬的，刚性的，严格的
beam [biːm]	n. 梁，横梁；v. 播送
transmission [trænz'mɪʃ(ə)n]	n. 播送，传送，传输，转播
differential [ˌdɪfə'renʃ(ə)l]	adj. 差别的；n. 差速器，差动装置
disengagement [dɪsɪn'geɪdʒm(ə)nt; dɪsen]	n. 脱离
retardation [ˌriːtɑː'deɪʃən]	n. 延迟，障碍物，制动
wankel engine	n. 转子发动机
mounted on	安装
power train	动力传动系
output shaft	输出轴
steer [stɪə]	n. 驾驶，转向
pedal ['pedl]	n. 踏板
master cylinder	主制动缸
brake drum	刹车鼓，制动鼓
tractive force	牵引力

⇨ Notes

[1] Finally, an electrical system is required to operate the cranking motor that starts the engine and to provide electrical energy to power engine accessories.

参考译文：最后，电子系统被用来控制启动发动机用的电动机和为发动机附件提供电能。

that 引导的定语从句修饰 motor；to operate 和 to provide 并列作目的状语。

[2] Such disengagement of the power train from the engine is essential while changing the gear ratio or while stopping the vehicle.

参考译文：当换挡或停车时，把传动系和发动机的连接断开是很必要的。

while 引导时间状语从句；gear ratio 表示"速比"。

[3] The major requirements in any steering mechanism are that it should be precise and easy to handle, and that the front wheels should have a tendency to return to the straight-ahead position after a turn.

参考译文：在任何转向机构中最主要的就是转向精确且容易控制，同时前轮在转向后又能自动回正。

两个 that 引导的是并列表语从句；Steering mechanism 表示"转向机制"。

[4] As the load on the vehicle and the vehicle speed has increased according to recent trends, in modern days, the importance of the brake system has also increased and power brakes are now being preferred.

参考译文：随着汽车负载和车速的增大，目前制动系统的重要性也在增大，并且现在人们更喜欢助力制动。

according to 表示"按照"的意思；as 引导时间状语从句。

Questions

1. What components does an automobile have?
2. What is the function of lubrication system?
3. What is the function of fuel system?
4. What types of suspension system are there?
5. What requirements should any steering mechanism meet?

Unit 25 CAD/CAM/CAPP

CAD

Computer aided design (CAD) can be defined as using computers to aid the engineering design process by means of effectively creating, modifying, or documenting the part's geometrical modeling. CAD is most commonly associated with the use of an interactive computer graphics system. The object of the engineering design is stored and represented in the form of geometric models. Geometric modeling is concerned with the use of a CAD system to develop a mathematical description of the geometry of an object. The mathematical description is called a model. There are three types of models (wire-frame models, surface models, and solid models, see Figure25-1) that are commonly used to represent a physical object. Wire-frame models, also called edge-vertex or stick-figure models, are the simplest method of modeling and are most commonly used to define computer models of parts. Surface models may be constructed using a large variety of surface features. Solid models are recorded in the computer result; it is possible to calculate mass properties of the parts, which is often required for engineering analysis such as finite element methods, kinematics or dynamic studies, and mass or heat transfer for interference checking [1].

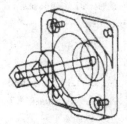 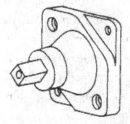 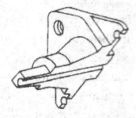

Figure 25-1   Three types of model

Models in CAD also can be classified as being two-dimensional models, two-and-half-dimensional models, or three-dimensional models. A 2-D model represents a flat part and a 3-D model (see Figure 25-2) provides representation of a generalized part shape. A 5/2-D model can be used to represent a part of constant section with no side-wall details. The major advantage of a 5/2-D model is that it gives certain a mount of 3-D information about a part without the need to create the database of a full 3-D model [2].

After a particular design alternative has been developed, some form of engineering analysis must often be performed as a part of the design process. The analysis may take the form of stress-strain calculations, heat transfer analysis, dynamic simulation etc. Some examples of the software typically offered on CAD systems are mass properties and Finite Element Method (FEM) analysis. Mass properties involve the computation of such features of a solid object as its volume, surface area, weight, and center of gravity. FEM analysis is available on most CAD systems to aid in heat transfer, stress-strain analysis, dynamic characteristics, and other engineering computations. Presently, many CAD systems can automatically generate the 2-D or 3-D FEM meshes which are essential to FEM analysis.

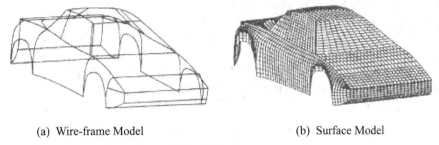

(a) Wire-frame Model (b) Surface Model

Figure 25-2 Three-dimension model

CAM

CAM can be defined as computer aided preparation manufacturing including decision-making, process and operational planning, software design techniques, artificial intelligence, manufacturing with different types of automation (NC machine, NC machine centers, NC machining cells, NC flexible manufacturing systems), and different types of realization (CNC single unit technology, DNC group technology).

The CAM covers group technology, manufacturing database, automated and tolerance. Figure 25-3 illustrates the general scope of CAM.

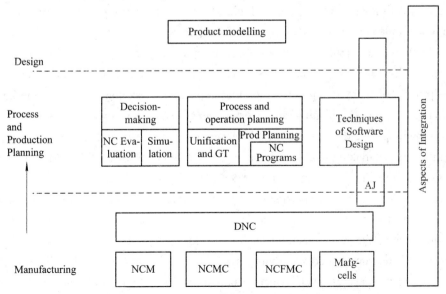

Figure 25-3 The general scope of CAM

When a design has frozen, manufacturing can begin. Computers have an important role to play in many aspects of production. Numerically controlled machine tools need a part program to define the components being made, computer techniques exist to assist, and in some cases virtually automate the generation of part programs. Modern shipbuilding fabricates structures from welded steel plates that are cut from a large steel sheet. Computer-controlled flame cutters are often used for this task and the computer is used to calculate the optimum layout of the components to minimize waste metal. Numerically controlled pipe-bending machines are able to operate directly from part programs generated by pipe-routing software.

Printed circuit board assembly can also be improved by computer methods. Quality is maintained by computer-controlled automatic test equipment that diagnoses faults in a particular board and rejects defective boards from the assembly line. Computers are used extensively to plot the artwork used to etch (蚀刻) printed circuit boards and also to produce part programs for NC drilling machines.

One of the most important manufacturing function is stock and production control. If the original design is done on a computer, obtaining lists of material requirements is straightforward. Standard computer data processing methods are employed to organize the work flow and order components when required.

CAPP

Computer aided process planning (CAPP) can be defined as the functions which use computers to assist the work of process planners. The levels of a assistance depend on the different strategies employed to implement the system. Lower level strategies only use computers for storage and retrieval of the data for the process plans which will be constructed manually by process planners as well as for supplying the data which will be used in the planner's new work. In comparison with lower level strategies, higher level strategies use computers to automatically generate process plans for some work pieces of simple geometrical shapes. Sometimes a process planner is required to input the data needed or to modify plans which do not fit specific production requirements well. The highest level strategy, which is the ultimate goal of CAPP, generates process plans by computer, which may replace process planners, when the knowledge and expertise of process planning and working experience have been incorporated into the computer programs. The database in a CAPP system based on the highest level strategy will be directly integrated with conjunctive systems, e.g. CAD and CAM. CAPP has been recognized as playing a key role in CIMS (Computer Integrated Manufacturing System).

More than 20 years have elapsed since the use of computers to assist process planning tasks was first proposed. Tremendous efforts have been made in the development of CAPP systems. For the time being, the research interests for development of CAPP systems are focused on intelligent and integrated process planning systems. For increasing the intelligence of CAPP systems, some new concepts, such as neural networks, fuzzy logic, and machine learning have been explored for the new generation of CAPP system. For increasing the integrability of CAPP system, feature based design, the roles of features, integrating process planning with scheduling, and integrating process planning with manufacturing resources planning have been focused on. This phenomenon is entitled concurrent or simultaneous engineering.

Since a process plan determines the methods, machines, sequences, fixturing and tools required in the fabrication and assembly of components, it is easy to see that process planning is one of the basic tasks to be performed in manufacturing systems. The task of carrying out the difficult and detailed process plans has traditionally been done by workers with a vast knowledge and understanding of the manufacturing process. Many of these skilled workers, now considered process planners, are either retired or close to retirement, with no qualified young process planners to take their place. An increasing shortage of process planners has been created. With the high pressure of serious of competition in the world market, integrated production has been pursued as a way for companies to survive and succeed. Automated process planning systems have been recognized as playing a key role in CIMS (Computer Integrated Manufacturing System). It is for reasons such as these that many companies look for computer aided process systems.

◇ **New Words and Expressions**

interactive [ɪntər'æktɪv]	adj. 交互式的
vertex ['vɜːteks]	n. 顶点，最高点
fabricate ['fæbrɪkeɪt]	v. 构成，伪造，虚构
straightforward [ˌstreɪt'fɔːwəd]	adj. 正直的，简单的，直截了当的
incorporate [ɪn'kɔːpəreɪt]	adj. 合并的，一体化的
phenomenon [fɪ'nɒmɪnən]	n. 现象
simultaneous [ˌsɪm(ə)l'teɪnɪəs]	adj. 同时的，同时发生的
wire-frame models	线框模型
surface models	表面模型
solid models	实体模型
stress-strain	应力—应变

⇨ **Notes**

[1] It is possible to calculate mass properties of the parts, which is often required for engineering analysis such as finite element methods, kinematics or dynamic studies, and mass or heat transfer for interference checking.

参考译文：计算汽车零部件的质量特性是可能的，也是工程分析中经常需要的，如有限元法、运动学或动力学研究、热传递干涉校验。

which 引导的非限定性定语从句修饰 mass properties；finite element methods 表示"有限元方法"；interference checking 表示"干涉校验"。

[2] The major advantage of a 5/2-D model is that it gives a certain amount of 3-D

information about a part without the need to create the database of a full 3-D model.

参考译文：5/2-D 模型的主要优点是，它给出了一定量有关某个零部件的三维信息，而不需要创建一个完整的三维模型数据库。

that 引导的表语从句作系动词 is 的表语；without 引导的介词短语修饰 part。

⊠ Questions

1. What is the definition of CAD?
2. What types of models are commonly used to represent a physical object?
3. How can models in CAD be classified?
4. What is the definition of CAM?
5. What is the definition of CAPP?

Reading Material: Components of a Robot System

The components of a robot system could be discussed either from a physical point of view or from a system point of view. Physically, we would divide the system into the robot, power system, and controller. Likewise, the robot itself could be partitioned anthropomorphically into base, shoulder, elbow, wrist, gripper, and tool. Most of these terms require little explanation.

Consequently, we will describe the components of a robot system from the point of view of information transfer, that is, what information or signal enters the component; what logical or arithmetic operation does the component perform; and what information or signal does the component produce? It is important to note that the same physical component may perform many different information processing operations (e.g. a central computer performs many different calculations on different data). Likewise, two physically separate components may perform identical information operations (e.g. the shoulder and elbow actuators both convert signals to motion in very similar ways).

Actuator associated with each joint on the robot is an actuator which causes that joint to move. Typical actuators are electric motors and hydraulic cylinders. Typically, a robot system will contain six actuators, since six are required for full control of position and orientation. Many robot applications do not require this full flexibility, and consequently, robots are often built with five or fewer actuators.

Sensor to Control Actuator. The computer must have information regarding the position and possibly the velocity are the actuator. In this context, the term position refers to a displacement from some arbitrary zero reference point for that actuator. For example, in the case of a rotary actuator, "position" would really the angular position and be measured in radians.

Many types of sensors can provide indications of position and velocity. The various types of sensors require different mechanisms for interfacing to the computer. In addition, the industrial use of the manipulator requires that the interface be protected from the harsh electrical environment of the factory. Sources of electrical noise such as arc welders and large motors can easily make a digital system useless unless care is taken in design and construction of the interface.

Computation. We could easily have labeled the computation module "computer" as most of the function to be described are typically performed by digital computers. However, many of the functions may be performed in dedicated custom hardware or networks of computers. We will, thus, discuss the computational component as if it were a simple computer, recognizing that the need for real-time control may require special equipment and that some of this equipment may even be analog, although the current trend is toward fully digital systems.

One further note: We will tend to avoid the use of the term microprocessor in this book and simply say computer, although many current robot manufacturers use one or more microprocessors in their systems.

The computation component performs the following operations:

Servo. Given the current position and/or velocity of an actuator, determine the appropriate drive signal to move that actuator toward its desired position. This operation must be performed for each actuator.

Kinematics. Given the current state of the actuators (position and velocity), determine the current state of the gripper. Conversely, given a desired state of the hand, determine the desired state for each actuator.

Dynamics. Given knowledge of the loads on the arm (inertia, friction, gravity, acceleration) use this information to adjust the servo operation to achieve better performance.

Workplace Sensor Analysis. Given knowledge of the task to be performed, determine appropriate robot motion commands. This may include analyzing a TV picture of the workplace or measuring and compensating for forces applied at the hand.

In addition to these easily identified components, there are also supervisory operations such as path planning and operator interaction.

◇ New Words and Expressions

physical ['fɪzɪk(ə)l]   a. 物质的，有形的，实际的，物理的，自然的

partition [pɑː'tɪʃ(ə)n]   n. & v. 划分，区分，分割，分离

anthropomorphically [ˌænθrəpəu'mɔːfikəli]   adv. 具有人的特点地，拟人地

gripper ['grɪpɚ]   n. 抓握器，夹持器，抓手

actuator ['æktjʊeɪtə]   n. 驱动器，执行机构

hydraulic [haɪ'drɔːlɪk; haɪ'drɒlɪk]   adj. 水力的，液压的；n. 液压传动装置

orientation [ɔrɪɛn'teiʃən]   n. 定向，朝向，定位，方位

sensor ['sensə]	n. 传感器，传感元件
indication [ˌɪndɪ'keɪʃ(ə)n]	n. 指示，显示，示数，读数
interface ['ɪntəfeɪs]	n. 界面，接口设备，连接装置
module ['mɒdjuːl]	n. 模量，模件，组件，模块
dedicated ['dedɪkeɪtɪd]	adj. 专用的
custom ['kʌstəm]	n. 习惯，顾客；adj. 定做的，定制的
realtime	实时，与发生的物理过程同步进行的计算
analog ['ænəlɒg]	n. 模拟量，模拟装置，模拟系统
servo ['sɜːvəʊ]	n. 伺服机构，伺服电机，伺服传动装置
energy ['enədʒɪ]	n. 能，能量
create [kriː'eɪt]	vt. 创造，引起，造成

◈ Questions

1. What does a robot system include from a physical point of view and the point of view of information transfer?
2. What operations does the computation component perform?
3. Why are robots often built with five or fewer actuators rather than 6 actuators?
4. How can a digital system be protected from sources of electrical noise?

PART 6　Electric Knowledge and Automobile

Unit 26 Numerical Control of Production Equipments (I)

Numerical control (NC) is a form of programmable automation in which the processing equipment is controlled by means of numbers, letters, and other symbols. The numbers, letters, and symbols are coded in an appropriate format to define a program of instructions for a particular workpart or job. When the job changes, the program of instructions is changed. The capability to change the program is what makes NC suitable for low-and medium-volume production. It is much easier to write new programs than to make major alterations of the processing equipment.

Basic Components of NC

A numerical control system consists of the following three basic components:
- Program of instructions
- Machine control unit
- Processing equipment

The general relationship among the three components is illustrated in Figure 26-1. The program is fed into the control unit, which directs the processing equipment accordingly.

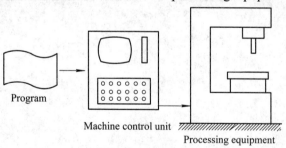

Figure 26-1 Basic components of an NC system

The program of instructions is the detailed step-by-step commands that direct the processing equipment. In its most common form, the commands refer to positions of a machine tool spindle with respect to the worktable on which the part is fixtured. More advanced instructions include selection of spindle speeds, cutting tools, and other functions. The program is coded on a suitable medium for submission to the worktable on which the part is fixture. More advanced instructions include medium for submission to the machine control unit. The program is coded on a suitable medium for submission to the machine control unit. The most common medium in use over the last several decades has been 1-in-wide punched tape. Coming into use more recently have been magnetic tape cassettes and floppy diskettes.

The machine control unit (MCU) consists of the electronics and control hardware that read and interpret the program of instruction and convert it into mechanical actions of the machine tool or other processing equipment.

The processing equipment is the third basic component of an NC system. It is the component that performs useful work. In the most common example of numerical control, one that performs machining operations, the processing equipment consists of the worktable and spindle as well as the motors and controls needed to drive them.

Types of Control System

There are two basic types of control systems in numerical control: point-to-point and contouring. In the point-to-point system, also called positioning, each axis of the machine is driven separately by lead screws and, depending on the type of operation, at different velocities. The machine moves initially at maximum velocity in order to reduce nonproductive time but decelerates as the tool reaches its numerically defined position. Thus is an operation such as drilling or punching, the positioning and cutting take place sequentially. After the hole is drilled or punched, the tool retracts, moves rapidly to another position, and repeats the operation. The path followed from one position to another is important in only one respect: the time required should be minimized for efficiency. Point-to-point systems are used mainly in drilling, punching, and straight milling operations.

In the contouring system, also known as the continuous path system, positioning and cutting operations are both along controlled paths but at different velocities. Because the tool cuts as it travels along a prescribed path, accurate control and synchronization of velocities and movements are important. The contouring system is used on lathes, milling machines, grinders, welding machinery, and machining centers.

Movement along the path, or interpolation, occurs incrementally, by one of several basic methods. In all interpolations, the path controlled is that of the center of rotation of the tool. Compensation for different tools, different diameter tools, or tool wear during machining can be made in the NC program.

There are a number of interpolation schemes that have been developed to deal with the various problems that are encountered in generating a smooth continuous path with a contouring –type NC system, they include:
- Linear interpolation
- Circular interpolation
- Helical interpolation
- Parabolic interpolation
- Cubic interpolation

Each of these interpolation procedures permits the programmer (or operator) to generate machine instructions for linear or curvilinear paths, using a relatively few input parameters. The interpolation module in the MCU performs the calculations and directs the tool along the path.

Linear interpolation is the most basic and is used when a straight-line path is to be generated in continuous-path NC. Two-axis and three-axis linear interpolation routines are sometimes distinguished in practice, but conceptually they are the same. The programmer is required to specify the beginning point and end point of the straight line, and the rate that is to be followed along the straight line. The interpolator computes the feed rates for each of the two (or three) axes in order to achieve the specified feed rate.

Linear interpolation for creating a circular path would be quite inappropriate because the programmer would be required to specify the line segments and their respective and points that are to be used to approximate the circle. Circular interpolation schemes have been developed that permit the programming of a path consisting of a circular by specifying the following parameters of the arc: the coordinates of its end points, the coordinates of its center, its radius, and the direction of the cutter along the arc. The tool path that is created consists of a series of straight-line segments, but the segments are calculated by the interpolation module rather than the programmer. The cutter is directed to move along each line segment one by one in order to generate the smooth circular path. A limitation of circular interpolation is that the plane in which the circular arc exists must be a plane defined by two axes of the NC system.

Helical interpolation combines the circular interpolation scheme for two axes described above with linear movement of a third axis. This permits the definition of a helical path in three-dimensional space.

Parabolic and cubic interpolation routines are used to provide approximations of free-form curves using higher-order equations. They generally require considerable computational power and are not as common as linear and circular interpolation. Their applications are concentrated in the automobile industry for fabricating dies for car body panels styled with free-form designs that cannot accurately and conveniently be approximated by combining linear and circular interpolations.

Programming for NC

A program for numerical control consists of a sequence of directions that causes an NC machine to carry out a certain operation, machining being the most commonly used process. Programming for NC may be done by an internal programming department, on the shop floor, or purchased from an outside source. Also, programming may be done manually or with computer assistance.

The program contains instructions and commands. Geometric instructions pertain to relative movements between the tool and the workplace. Geometric instructions pertain to relative speeds, feeds, tools, and so on. Travel instructions pertain to the type of interpolation and slow or rapid movements of the tool or worktable. Switching commands pertain to on/off position for coolant supplies, spindle rotation, direction of spindle rotation, tool changes, workplace feeding, clamping, and so on.

(1) Manual Programming. Manual part programming consists of first calculating dimensional relationships of the tool, workplace, and worktable, based on the engineering drawings of the part, and manufacturing operations to be performed and their sequence. A program sheet is cutting tools, spindle speeds, feeds, depth of cut, cutting fluids, power, and tool or workplace relative positions and movements. Based on this information, the part program is prepared. Usually a paper tape is first prepared for trying out and debugging the program. Depending on how often it is to be used, the tape may be made of more durable mylar.

Manual programming can be done by someone knowledgeable about the particular process and able to understand, read, and change part programs. Because they are familiar with machine tools and process capabilities, skilled machinists can do manual programming with some training in programming. However, the work is tedious, time consuming, and uneconomical—and is used mostly in simple point-to-point applications.

(2) Computer-Aided Programming. Computer-aided part programming involves special symbolic programming languages that determine the coordinate points of corners, edges, and surfaces of the part. Programming language is the means of communicating with the computer and involves the use of symbolic characters. The programmer describes the component to be processed in this language, and the computer converts it to commands for the NC machine [1]. Several languages have various features and applications are commercially available. The first language that used English-like statements was developed in the late 1950 s and is called APT (for Automatically Programmed Tools). This language, in its various expanded forms, is still the most widely used for both point-to-point and continuous-path programming.

Computer-aided part programming has the following significant advantages over manual method:

- Relatively easy to use symbolic language.
- Reduced programming time. Programming is capable of accommodating a large amount of data concerning machine characteristics and process variables, such as power, speeds, feed, tool shape, compensation for tool shape changes, tool wear, deflections, and coolant use.
- Reduced possibility of human error, which can occur in manual programming.
- Capability of simple changeover of machining sequence or from machine to machine.
- Lower cost because less time is required for programming.

Selection of a particular NC programming language depends on the following factors:

(a) Level of expertise of the personnel in the manufacturing facility.
(b) Complexity of the part.
(c) Type of equipment and computers available.
(d) Time and costs involved in programming.

Because numerical control involves the insertion of data concerning workplace materials and processing parameters, programming must be done by operators or programmers who are knowledgeable about the relevant aspects of the manufacturing processes being used. Before production begins, programs should be verified, either by viewing a simulation of the process on a CRT screen or by making the part from an inexpensive material, such as aluminum, wood, or plastic, rather than the material specified for the finished part.

NC Part Programming Languages

Probably over 100 NC part programming languages have been developed since the initial MIT research [2] on NC programming systems in 1956. Most of the languages were developed to serve particular needs and machines and have not survived the test of time. However, a good number of languages are still in use today. In this subsection we review some of those which are generally considered important.

APT (Automatically Programmed Tools). The APT language was the product of the MIT development work on NC programming systems. Its development began in June 1956, and it was first used in production around 1959. Today it is the most widely used language in the United States. Although first intended as a contouring language, modern versions of APT can be used for both positioning and continuous-path programming in up to five axes.

AUTOSPOT (Automatic System for Positioning Tools). This was developed by IBM and first introduced in 1962 for PTP programming. Today's version of AUTOSPOT can be used for contouring as well.

COMPACT 11. This is a package available from Manufacturing Data Systems, Inc. (MDSI), a firm based in Ann Arbor, Michigan. The NC language is similar to SPLIT in many of its features. MDSI leases the COMPACT 11 system to its users on a time-sharing basis. The part programmer uses a remote terminal to feed the program into the MDSI computers, which in turn produces the NC tape.

ADAPT (Adaptation of APT). Several part programming languages are based directly on the APT program. One of these is ADAPT, which was developed by IBM under Air Force contract. It was intended to provide many of the features of APT but to utilize a significantly smaller computer. ADAPT is not as powerful as APT, but can be used to program for both positioning and contouring jobs.

EXAPT (Extended Subset of APT) this was developed in Germany starting around 1964 and is based on the APT language. There are three versions: EXAPT I —designed for positioning (drilling and also straight-cut milling), EXAPT II—designed for turning, and EXAPT III —designed for limited contouring operations. One of the important features of EXAPT is that it attempts to compute optimum feeds and speeds automatically.

APT is not only an NC language; it is also the computer program that performs the calculations to generate cutter positions based on APT statements.

There are four types of statements in the APT languages:

• Geometry statements. These define the geometric elements that comprise the work part . They are also sometimes called definition statements.

• Motion statements. These are used to describe the path taken by the cutting tool.

• Postprocessor statements. These apply to the specific machine tool and control system. They are used to specify feeds and speeds and to actuate other features of the machine.

• Auxiliary statements. These are miscellaneous statements used to identify the part, tool, tolerances, and so on.

◇ New Words and Expressions

geometry [dʒi'ɑ:mətri]   n. 几何学
postprocessor [poʊstp'roʊsesə]   n. 后处理程序
package ['pækɪdʒ]   n. 包裹，包装袋
workplace ['wɜ:rkpleɪs]   n. 工作场所，车间

⇨ Notes

[1] The programmer describes the component to be processed in this languages, and the computer converts it to commands for the NC machine.

参考译文：编程员用这种语言描述加工零件，而由计算机将零件程序转换为数控机床的执行指令。

The component to be processed——被加工零件。

[2] the initial MIT research.

参考译文：最初的麻省理工学院的研究。

MIT——Massachusetts Institute of Technology (麻省理工学院)。

✗ Questions

1. What is the meaning of NC?
2. What are the basic components of NC systems?
3. What are the NC programming languages?
4. What are the features of NC programs?

Reading Material: Numerical Control of Production Equipments (II)

The most common application of numerical control is for machine tool control. This was the first application of NC and is today the most important commercially. In this section we discuss the machine tool application of NC with emphasis on metal machining.

Machine Tool Technology for NC

Each of the five machining processes is carried out on a machine tool designed to perform that process. Turning is performed on a lathe, drilling is done on a drill press, milling on a milling machining, and so on. There are several different types of grinding operations with a corresponding variety of machine to perform them. Numerical control machine tools have been designed for nearly all of the machining processes. The list includes:

1) Drill presses.
2) Milling machines, vertical spindle and horizontal spindle.
3) Turning machines, both horizontal axis and vertical axis.
4) Horizontal and vertical boring mills.
5) Profiling and contouring mills.
6) Surface grinders and cylindrical grinders.

In addition to the machining process, NC machine tools have also been developed for other metal working processes. These machines include:

1) Punch presses for sheet metal hole punching.
2) Presses for sheet metal bending.

The introduction of numerical control has had a pronounced influence on the design and operation of machine tools. One of the effects of NC has been that the proportion of time spent by the machine cutting metal under program control is significantly greater than with manually operated machines. This causes certain components, such as the spindle, drive gears, and feed screws, to wear more rapidly. These components must be designed to last longer on NC machines. Second, the addition of the electronic control unit has increased the cost of the machine, therefore requiring higher equipment utilization. Instead of running the machine on only one shift, which was the convention with manually operated machines, NC machines are often operated two or even three shifts to obtain the required payback. Also, the NC machines are designed to reduce the time consumed by the nonprocessing elements in the operation cycle, such as loading and unloading the workpart, and changing tools. Third, the increasing cost of labor has altered the relative roles of the operator and the machine tool. Consider the role of the human operator. Instead of the operator have been reduced to part loading and unloading, tool changing, chip clearing, and the like. In this way, one operator can often run two or three automatic machines. The role and functions of the machine tool have also changed. NC machines are designed to be highly automatic and capable of combining several operations in one setup that formerly required several different machines. These changes are best exemplified by a new type of machine that did not exist prior to the advent and development of numerical control: the machining center.

The machining center, developed in the late 1950s, is a machining tool capable of performing several different machining operations on a workpart in one setup under program control. The machining center is capable of milling, drilling, reaming, tapping, boring, facing, and similar operations. In addition, the features that typically characterize the NC machining center include the following:

1) Automatic tool-changing capability. A variety of machining operations means that a variety of tools is required. The tools are contained on the machine in a tool magazine or drum. When a tool needs to be changed, the tool drum rotates to the proper position, and an automatic tool changing mechanism, operating under program control, exchanges the tool in the spindle and the tool in the drum.

2) Automatic workpart positioning. Most machining centers have the capability to rotate the job relative to the spindle, thereby permitting the cutting tool to access four surfaces of the part.

3) Pallet shuttle. Another feature is that the machining centers have two (or more) separate pallets that can be presented to the cutting tool. While machining is being performed with one pallet in position in front of the tool, the other pallet is in a safe location away from the spindle. In this way, the operator can be unloading the finished part from the prior cycle and fixturing the raw workpart for the next cycle while machining is being performed on the current workpiece.

Machining centers are classified as vertical or horizontal. The descriptor refers to the orientation of the machine tool spindle. A vertical machining center has its spindle on a vertical axis relative to the worktable, and a horizontal machining center has its spindle on a vertical axis. This distinction generally results in a difference in the type of work that is performed on the machine. A vertical machining center is typically used for flat work that requires tool access form be achieved on the sides of the cube. An example of an NC horizontal machining center, capable of many of the features described above, is shown in Figure 26-2.

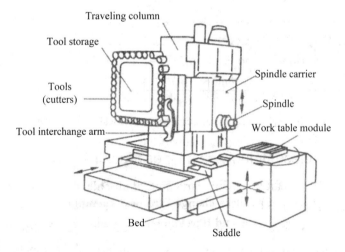

Figure 26-2 Schematic illustration of a horizontal spindle machining center, equipped with an automatic tool changer

The success of the machining center has resulted in the development of similar machine tools for other metalworking processes. One example is the turning center, designed as a highly automated and versatile machine tool for performing turning, facing, drilling, threading, and related operations.

DNC and CNC

The development of numerical control was a significant achievement in batch and job shop manufacturing, from both a technological and a commercial viewpoint. There have been two enhancements and extensions of NC technology, including:

1) Direct numerical control.
2) Computer numerical control.

Direct Numerical Control

Direct numerical control can be defined as a manufacturing system in which a number of machines are controlled by a computer through direct connection and in real time. The tape reader is omitted in DNC, thus relieving the system of its least reliable component. Instead of using the tape reader, the part program is transmitted to the machine tool directly from the computer memory. In principle, one computer can be used to control more than 100 separated machines. (One commercial DNC system during the 1970s boasted a control capability of up to 256 machine tools.) The DNC computer is designed to provide instructions to each machine tool on demand. When the machine needs control commands, they are communicated to it immediately. Figure 26-3 illustrates the general DNC configuration. The system consists of four components.

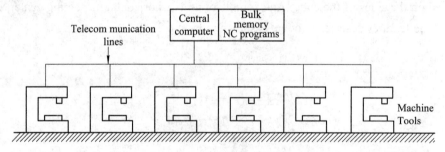

Figure26-3　General Configuration of a direct numerical control (DNC) system

Telecommunication Lines Machine Tools

The computer calls the part program instructions from bulk storage and sends them to the individual machines as the need arises. It also receives data back from the machines. This two-way information flow occurs in real time, which means that each machine's requests for instructions must be satisfied almost instantaneously. Similarly, the computer must always be ready to receive information from the machines and to respond accordingly. The remarkable feature of the DNC system is that the computer is servicing a large number of separate machine tools, all in real time. Depending on the number of machines and the computational requirements that are imposed on the computer, it is sometimes necessary to make use of satellite computers, as shown in Figure 26-4. These satellites are smaller computers, and they serve to take some of the burden off the larger central computer. Groups of part program instructions are received from the central computer and stored in buffers; they are then dispensed to the individual machines as required. Feedback data from the machines are also stored in the satellite's buffer before being collected at the central computer.

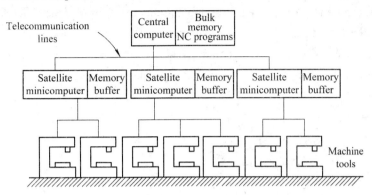

Figure26-4 DNC in hierarchical configuration using satellite computer

Computer Numerical Control

Since the introduction of DNC, there have been dramatic advances in computer technology. The physical size and cost of a digital computer has been significantly reduced at the same time that its computational capabilities have been substantially increased. In numerical control, the result of these advances has been that the large hard-wired MCUs of conventional NC have been replaced by control units based on the digital computer. Initially, minicomputers were replaced by today's microcomputers.

Computer numerical control is an NC system using dedicated microcomputer as the machine control unit. Because a digital computer is used in both CNC and DNC, it is appropriate to distinguish between the two types of system. There are three principal differences:

1) DNC computers distribute instruction data to, and collect data from, a large number of machines. CNC computers control only one machine, or a small number of machines.

2) DNC computers occupy a location that is typically remote from the machines under their control. CNC computers are located very near their machine tools.

3) DNC software is developed not only to control individual pieces of production equipment, but also to serve as part of a management information system in the manufacturing sector of the firm. CNC software is developed to augment the capabilities of a particular machine tool.

The general configuration of a computer numerical control system is pictured in Figure 26-5. As illustrated in the diagram, the controller has a tape reader for initial entry of a part program. In this regard, the outward appearance of a CNC system is similar to that of a conventional NC machine. However, the way in which the program is used in CNC is different. With a conventional NC system, the punched tape is cycled through the tape reader for each workpart in the batch. The MCU reads in a block of instructions on the tape, executing that block before proceeding to the next block. In CNC, the entire program is entered once and stored in computer memory. The machining cycle for each part is controlled by the program contained in memory rather than from the tape itself.

Control algorithms contained in the computer convert the part program instructions into actions of the machine tool (or other processing equipment). Certain functions are carried out by hard-wired components in the MCU. For example, circular interpolation calculations are often performed by hard-wired circuits rather than by stored program. Also, a hardware interface is required to make the connections with the machine tool servo-systems.

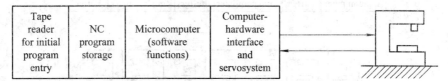

Figure 26-5　General configuration of a computer numerical control (CNC) system

CNC Control Features

In the previous chapter, the important features and functions of the machine control unit in numerical control were described. CNC has made possible additional features beyond what is normally found in a conventional hard-wired MCU. Some of these features include the following:

1) Storage of more than one part program. With improvements in computer technology, many of the newer CNC controllers have a large enough capacity to store more than a single program. This translates into the capability to store either one very large program or several small and medium-sized programs.

2) Use of diskettes. There is a growing use of floppy disks for part programs in manufacturing. Diskette is the approximate equivalent of 8000 ft. of punched tape. Because of this more modern storage technology, many CNC controllers have the optional capability to read in programs stored on disks as well as punched tape.

3) Program editing at the machine tool site. To deal with the mistakes in part programming, CNC systems permit the program to be edited while it is in computer memory. Hence, the process of testing and correcting the program can be done entirely at the machine site rather than returning to the programming office in the shop to make the tape corrections. In addition to part program corrections, editing can also be done to optimize the cutting conditions of the machining cycle. After correcting and optimizing the program, a tape punch can be connected to the CNC controller in order to obtain a revised version of the tape for future use.

4) Fixed cycles and programming subroutines. The increased memory capacity and the ability to program the control computer in CNC provides the opportunity to store frequently used machining cycles in memory that can be called by the part program. Instead of writing the instructions for the particular cycle into every program, a code is written into the program to indicate that the cycle should be executed. Some of these cycles require the definition of certain

parameters in order to execute. An example is a bolt hole circle, in which the diameter of the bolt circle, the spacing of the bolt holes, and other parameters must be specified. In other cases, the particular machining cycle used by the shop would not require parameter definition.

5) Interpolation. Some of the interpolation schemes are normally executed only on a CNC system because of the computational requirements. Linear and circular interpolations are often hard-wired into the control unit. Helical, parabolic, and cubic interpolations are usually executed in a stored program algorithm.

6) Positioning features for setup. Setting up the machine tool for a certain job involves installing and aligning the fixture on the machine tool table. This must be accomplished so that the machine axes are aligned with the workpart. The alignment task can be facilitated using certain features that are made possible by software options in a CNC system. Position set is one of these features. With position set, the operator is not required to position the fixture by using a target point or set of target points on the work or fixture.

7) Cutter length compensation. This is similar to the preceding feature but applies to tool length and diameter. In older-style controls, the cutter dimensions had to be set very precisely in order to agree with the tool path defined in the part program. Other methods for ensuring accurate tool path definition have been incorporated into newer CNC controls. One method involves manually entering the actual tool dimensions into the MCU. These actual dimensions may differ from those originally programmed. Compensations are then automatically made in the computed tool path. Another more recent innovation is to use a tool length sensor built into the machine. In this method, the cutter is mounted in the spindle and brought into contact with the sensor to measure its length. This measured value is then used to correct the programmed tool path.

8) Diagnostics. Many modern CNC machines process an on-line diagnostics capability which monitors certain aspects of the machine tool and MCU operation to detect malfunctions or signs of impending malfunctions. When a malfunction is detected, or measurements indicate that a breakdown is about to occur, a message is displayed on the controller's CRT monitor. Depending on the seriousness of the malfunction, the system can be stopped or maintenance can be scheduled for a nonproduction shift. Another use of the diagnostics capability is to help the repair crew determine the reason for a break down of the machine tool. One of the biggest problems when a machine failure occurs is often in diagnosing the reason for the breakdown. By monitoring and analyzing its own operation, the system can determine and communicate the reason for the failure.

9) Communications interface. With the trend toward interfacing and net working in plants today, most modern CNC controllers are equipped with a standard communications interface to allow the particular machine tool to be linked to other computers and computer-driven devices.

◇ *New Words and Expressions*

diagnostics [ˌdaɪəɡˈnɒstɪks]　　　　n. 诊断学

interpolation [ɪn,tɚpə'leʃən]   n. 篡改，添写，插补
servosystem ['sɜːvoʊsɪstəm]   n. 伺服系统
commercially [kə'məːʃəlɪ]   adv. 商业上，通商上
numerical control   数字控制

Questions

1. What machining processes are needed in the design of numerical control machine tools?
2. What are the features of the typical NC machining center?
3. What are the main differences between DNC computers and CNC computers?
4. What can the machining center do?
5. What are the effects of NC?

Unit 27 Basic Electricity and Magnetism

To understand the theory of how an electric current flow, you must understand something about the structure of matter. Matter is made up of atoms. Atoms are made up of protons, neutrons, and electrons. Protons and neutrons are located at the center (or nucleus) of the atom. Protons have a positive charge. Neutrons have no charge and have little or no effect as far as electrical characteristics are concerned. Electrons have a negative charge and travel around the nucleus in orbits. The number of electrons in an atom is the same as the number of protons. Electrons in the same orbit is the same distance from the nucleus but do not follow the same orbital paths (Figure 27-1).

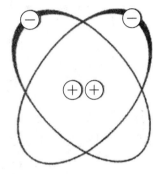

Figure 27-1 Orbital paths of electrons

The hydrogen atom is a simple atom to illustrate because it has only one proton and one electron (Figure 27-2). Not all atoms are as simple as the hydrogen atom. Most wirings used to conduct an electrical currents made of copper. Figure 27-3 illustrates a copper atom, which has 29 protons and 29 electrons. Some electron orbits are farther away from the nucleus than others. As can be seen, 2 travel in an inner orbit, 8 in the next, 18 in the next, and 1 in the outer orbit. It is this single electron in the outer orbit that makes copper a good conductor[1]. When sufficient energy or force is applied to an atom, the outer electron (or electrons) becomes free and moves. If it leaves the atom, the atom will contain more protons than electrons. Protons have a positive charge. This means that this atom will have a positive charge[2], Figure 27-4(a). The atom the electron joins will contain more electrons than protons, so it will have a negative charge, Figure 27-4(b). Like charges repel each other, and unlike charges attract each other. An electron in an atom with a surplus of electrons (negative charges) will be attracted to an atom with a shortage of electrons (positive charge). An electron entering an orbit with a surplus of electrons will tend to repel an electron already there and cause it to become a free electron.

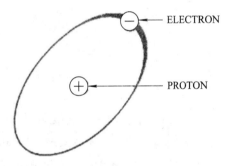

Figure 27-2 Hydrogen atom with one electron and one proton

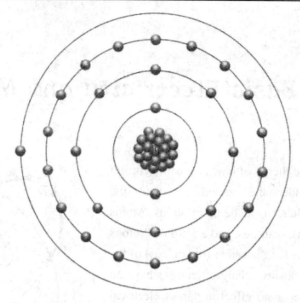

Figure 27-3 Copper atom with 29 protons and 29 electrons

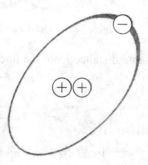 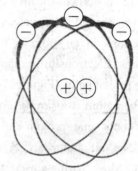

(a) An atom has two protons and one electron;   (b) An atom has two protons and three electrons

Figure 27-4 Atom shortage of electrons and its charge

Good conductors are those with few electrons in the outer orbit. Three common metals—copper, silver, and gold—are good conductors, and each has one electron in the outer orbit. These are considered to be free electrons because they move easily from one atom to another.

Atoms with several electrons in the outer orbit are poor conductors. These electrons are difficult to free, and materials made with these atoms are considered to be insulators. Glass, rubber, and plastic are examples of good insulators.

◇ *New Words and Expressions*

  circuit ['sə:kit]   n. 电路
  magnetism ['mægnɪˌtɪzəm] n. 磁，磁力，磁学
  matter ['mætə]   n. 物质
  atom ['ætəm]   n. 原子
  proton ['prəutɒn]   n. [核]质子

Unit 27　Basic Electricity and Magnetism

neutron ['nju:trɒn]	n. 中子
electron [i'lektrɔn]	n. 电子
nucleus ['nju:kliəs]	n. 核子
positive ['pɒzətɪv]	adj. [电]阳的
charge [tʃɑ:dʒ]	n. 电荷
characteristic [ˌkærəktə'rɪstɪk]	adj. & n. 特有的，表示特性的；特性，特征
negative ['negətiv]	adj. 负的，阴性的
orbit ['ɔ:bit]	n. 轨道
illustrate ['ɪləstreɪt]	vt. & vi. 举例说明，图解，阐明；举例
figure ['fɪgə(r)]	n. 图形
hydrogen ['haɪdrədʒən]	n. 氢
conduct [kən'dʌkt]	n. & v. 行为，引导，为人，传导
wiring ['waɪərɪŋ]	n. 配线
inner ['ɪnə(r)]	adj. & n. 内部的，里面的，内心的；内部
outer ['autə]	adj. 外部的，外面的，远离中心的
sufficient [sə'fɪʃnt]	adj. 充分的，足够的
repel [ri'pel]	vt. 击退，抵制，使厌恶，使不愉快
attract [ə'trækt]	vt. & vi. 吸引；有吸引力，引起注意
surplus ['sə:pləs]	n. & adj. & vt. 剩余，过剩的，剩余的；转让，卖掉
silver ['silvə]	n. & vt. 银，银子；镀银
insulator ['ɪnsjuleɪtə(r)]	n. 绝缘体，绝热体
be made of	由……组成
be located at	位于，坐落在……
as far as	就……来说，远到，直到，到……为止
as far as...be concerned	就……而言
the number of	……数字
the same as	与……相同
as...as	像……一样

⇨ Notes

[1]　It is this single electron in the outer orbit that makes copper a good conductor.
参考译文：正是这个在外层轨道上的单个电子使铜成为良导体。
由 It is...that... 引导的强调句，可译为：正是……才……。

[2]　This means that this atom will have a positive charge.
参考译文：这就意味着这个原子具有正电荷。
由 that 引导一个宾语从句，做 means 的宾语。

Questions

1. What is matter made of?
2. What are the components of an atom?
3. What are the features of good conductors?
4. Why is a hydrogen atom very simple?

Reading Material: Direct-current Circuits

The basic rules governing direct-current circuits, together with an introduction to the concepts of work, energy, and power, is discussed in this chapter. Line drop, line loss, wire gauges, resistivity, and temperature coefficient of resistance, all items relating directly to electric circuits, are also included. There are two basic ways of connecting two or more pieces of electric apparatus: they may be connected in series or in parallel. When electric devices are connected end-to-end or in tandem to form a single continuous circuit, they are said to be connected in series. The three resistances R1, R2 and R3 in Figure 27-5 are connected in series. Note that in Figure27-5 there is only one path over which current may flow.

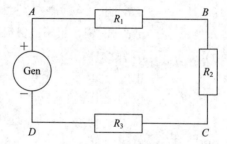

Figure 27-5　The resistances R1, R2 and R3 are connected in series

When the apparatus in connected so that there is a divided path over which current may flow two or more alternative routes between two points in a circuit, the arrangement is called a parallel circuit[1]. This combination shown in Figure 27-6 is also known as the multiple or shunt connection.

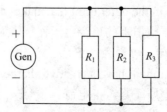

Figure 27-6　The resistances R1, R2 and R3 are connected in parallel

Other connections are combinations or variations of the two basic circuits. For example, in Figure27-7, the parallel combination of R1 and R2 is in series with R3. In Figure 27-8, the series combination of R1 and R2 is in parallel with R3. As still more apparatus is connected into a circuit, the combinations become more complex.

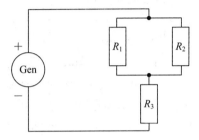

Figure 27-7 The parallel combination of R1 and R2 are connected in series with R3

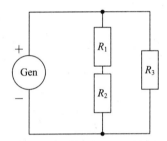

Figure 27-8 The series combination of R1 and R2 is connected in parallel with R3

Series Circuits: Current Relations

The same amount of current must flow in every part of a series circuit. This should be evident from a diagram of a series circuit. For example, in Figure 27-5 the current that flows out of the positive terminal of the generator must flow successively through resistances R1, R2 and R3 before returning to the negative terminal of the generator since there is only one path over which the current may flow. This is true regardless of the values of the several resistances in series. This may be easily verified experimentally by inserting an ammeter at several different point in a series circuit, such as at the points A, B, C or D in Figure 27-5. It will be found that the amount of current flowing at any of these points is the same. The current is the same in all parts of a series circuit.

This rule may be expressed mathematically by the expression

$$I = I_1 = I_2 = I_3 = \cdots \tag{27-1}$$

Where I is the current supplied by the generator, and I_1, I_2 and I_3 are the currents in the several parts of the circuit.

This does not imply that the current of a series circuit cannot be changed by altering the circuit. A change in either the applied voltage or the resistance of the circuit will change the value of the current flowing. However, for any given value of circuit resistance and applied voltage, the same current must flow in every part of the circuit.

Series Circuits: Voltage Relations

Water pressure is required to cause water to flow through a pipe. More pressure is required to force water to flow at a given rate through a small pipe than through a large pipe since more resistance is offered to the flow in the small pipe.

Likewise, in an electric circuit an electric pressure or voltage is necessary to cause a current to flow. The greater the resistance of a circuit, the greater the voltage must be to cause a given current to flow through that circuit. In Figure 27-9, voltmeters V1, V2 and V3 are connected across R1, R2 and R3, respectively. These voltmeters indicate the voltage required to cause the current of 2A, as indicated by the ammeter A, to flow through each of the three resistors. Voltmeter V indicates the voltage at the generator terminals, or the total voltage applied to the three resistors, which in Figure 27-9 is 72V.

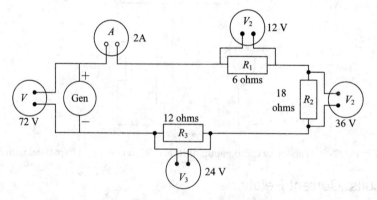

Figure 27-9 The sum of the voltages as indicated by voltmeters V1,V2 and V3 is equal to the voltage indicated by voltmeters

Voltmeter V1 indicates a voltage of 12 V cross R1. This means that 12 of the total of 72V is used in causing the current to flow through R1. Stated another way, there is a drop in voltage of 12 V across R1. Voltmeter V2 indicates 36V, or three times the voltage required across R1.This is to be expected since R2 has three times the resistance of R1 and the same current of 2A is being caused to flow through both resistors. The voltage across R3 is only 24V since R3 is only twice as large as R1. In general, it may be stated that the voltage required to cause a current to flow in a dc circuit is directly proportional to the resistance of the circuit; that is, the higher the resistance, the higher the voltage must be.

Note that in Figure 27-9 the sum of the voltage drops across the three resistors is equal to the voltage applied to the circuit (12 + 36 + 24 = 72).

The entire applied voltage is used in causing the current to flow through the three resistors. In a series circuit, the sum of the voltage across the several parts is equal to the total voltage applied to the circuit.

Stated mathematically:

$$E = E_1 + E_2 + E_3 + \cdots$$

Where E is voltage applied to the circuit and E_1, E_2 and E_3 are the voltages across the several parts.

◇ New Words and Expressions

govern ['gʌvən]	v. 管理，支配
gauge [gedʒ]	n. 标准，规则
apparatus [ˌæpə'reɪtəs]	n. 仪器，装置
evident ['evidənt]	adj. 明显的，明白的
regardless [rɪ'gɑːdləs]	adj. 不注意的
verify ['verifai]	vt. 证明检验，证实
rule [ruːl]	n. 规则，规律，定律
imply [ɪm'plaɪ]	vt. 暗示，意味着
cause [kɔːz]	vt. 原因，导致，引起
regardless of	不管……
stated another way	换句话说
as long as	只要是
work with	同……打交道

⇨ Notes

[1] When the apparatus in connected so that there is a divided path over which current may flow two or more alternative routes between two points in a circuit the arrangement is called a parallel circuit.

参考译文：当设备的接法是使得有一组分路可供电流通过，即电路的两点间有两条或多条可能的通路时，这种布置就称为并联电路。

本句是主从复合句。When 引导条件状语从句，so that there is a divided path 为结果状语从句，over which 引导定语从句以修饰 path。

⊠ Questions

1. What is the meaning of direct current?
2. What are series circuits?
3. What are parallel circuits?
4. What is the function of a resistance?

Unit 28　Electric Powers

The unit of electric power in both the U.S.Customary and the metric systems is the watt. In SI units, one watt is defined as being equal to work being done at the rate of one joule per second. The watt is also defined as the energy expended or the work done per second by an unvarying current of one ampere flowing under a pressure of one volt, or

$$P = IE \qquad (28\text{-}1)$$

where P = power, W; I = current, A; E = voltage, V

Example 28-1

What is the power used by an electric lamp that draws 2.5 A from a 120 V line?

$$P = IE$$
$$P = 2.5 \times 120 = 300 \text{ W}$$

The fact that the watt is a unit of power or a unit of a rate of doing work cannot be emphasized too strongly [1]. It will be remembered that current in amperes is a rate of flow of electricity or is equal to the number of coulombs per second.

The power formula may then be written

$$\text{Power in watts} = \text{coulombs/second} \times \text{volts}$$

In other words, the watt is a measure of how fast a quantity of electricity is being moved through a difference in potential. Since by Ohm's law $E=IR$, this value of E may be substituted in Eq. (28-1) to obtain another useful power formula

$$P = IE = I \times IR, \text{ or } P = I^2 R \qquad (28\text{-}2)$$

Example 28-2

What is the power used in a 60-ohm generator field rheostat when the field current is 2A?

$$P = I^2 R$$
$$P = 2^2 \times 60 = 4 \times 60 = 240 \text{ W}$$

A third power formula may be derived from the fact that $I = E/R$ by Ohm's law. Substituting in Eq. (28-1)

$$P = IE = E \times E/R \qquad (28\text{-}3)$$

or

$$P = E^2 /R$$

Example 28-3

What is the power used by a 15-ohm electric heater when a voltage of 120 V is applied?

$$P = E^2/R$$
$$P = 120^2/15 = 14\,400/15 = 960 \text{ W}$$

As illustrated in the above examples, Eq.(28-1) is used to find the power in a circuit when the current and voltage are known, Eq. (28-2) when the current and resistance are known, and Eq. (28-3) when the voltage and resistance are known.

Since the watt is a small unit, a larger unit, the kilowatt (kW) is often used instead. One kilowatt is equal to 1000watts.

Calculations concerning electric machinery often involve both the electric unit or power (watt) and the mechanical unit (horsepower). One horsepower is equal to 746 watts. Therefore, to change power in watt to power in horsepower, it is necessary to divide the number of watt by 746.

Example 28-4

The input to motor is 20 kW. What is the horsepower input?

$$20 \text{ kW} = 20 \times 1000 = 20\,000 \text{ W}$$
$$\text{Horsepower} = \text{watts}/746 = 20\,000/746 = 26.8 hp$$

For most purposes, the relation between the horsepower and the kilowatt may be taken as

$$1 hp = 3/4 \text{ kW (approximately)}$$

Power is a measure of how fast work is being done or of how fast energy is being expended, that is,

$$\text{Power} = \text{work or energy}/\text{time}$$

Thus, the energy used by an electric device is the rate at which the energy is being used (the power) multiplied by the time during which the device is in use. When power is measured in watts and time hours, then

$$\text{Power} \times \text{time} = \text{energy}$$

or

$$\text{Watts} \times \text{hours} = \text{watthours}$$

the watt-hour (Wh) being the energy expended when 1 watt is used for 1 hour.

The watthour is a relatively small unit; the kilowatthour being used much more extensively in com.One kilowatthour is equal to 1000 watthours.

Example 28-5

How much energy used by a 1500W heater in 8h?

$$\text{Energy} = \text{Power} \times \text{time}$$
$$= 1500 \times 8 = 12\,000 \text{ W} \cdot \text{h}$$
$$= 12\,000/1000 = 12 \text{ kW} \cdot \text{h}$$

If power is measured in watts and time in seconds, then

$$\text{Power} \times \text{time} = \text{energy}$$

and

$$\text{Watts} \times \text{seconds} = \text{watt-seconds}$$

The watt-second is called a joule, which is the SI unit for electric as well as mechanical energy.

Since there are 3,600 seconds in an hour and 1,000 watts in a kilowatt, one kilowatthour is equal to 3,600,000 joules or 3.6 megajoules (3.6 MJ).

Example 28-6

How much energy in joules, megajoules, and kilowatthours is used by a 100-watt lamp in 12 hour?

$$\text{Energy in joules} = \text{watts} \times \text{seconds} = 100 \times 12 \times 3600 = 4\,320\,000 \text{ J} = 4.32 \text{ MJ}$$
$$\text{Energy in kW} \cdot \text{h} = \text{watts} \times \text{hours}/1000 = 100 \times 12/1000 = 1.2 \text{ kW} \cdot \text{h}$$
$$1.2 \text{ kW} \cdot \text{h} \times 3.6 = 4.32 \text{ MJ (chech)}$$

Power is the rate of expending energy just as speed is a rate of motion. If the average speed of an automobile is known for a given time, the distance traveled is the average speed multiplied by the time traveled. Likewise, if the average power required by an electric motor for a given time is known, the energy used by the motor is the average power multiplied by the time the motor is used [2]. The reader should make sure that he understands the difference between power and energy. Power is the rate of expending energy or of doing work, just as speed is a rate of motion.

The commonly used electrical units of energy, work, and power are summarized in the following:

The USCS unit of work or energy = watthour(W·h).
The SI unit of work or energy = joule (J).
One watthour = 3600 joules.
One kilowatthour = 3.6 megajoules (MJ).
The USCS unit of power = watt (W).
The SI unit of power = watt(W).

◇ New Words and Expressions

watt [wɔt]	n.	瓦特，瓦 (电功单位)
customary ['kʌstəməri]	adj.	习惯的，惯常的
expend [ɪk'spend]	v.	花，耗费，消耗
unvarying [ʌn'veərɪɪŋ]	adj.	不变的，经常的，恒久的
ampere ['æmpeə(r)]	n.	安培
emphasize ['emfəsaiz]	v.	强调，使突出
coulomb ['ku:lɒm]	n.	库仑
formula ['fɔ:mjələ]	n.	公式
potential [pə'tenʃəl]	n.	势能，电位；可能
rheostat ['ri:əˌstæt]	n.	变阻器

heater ['hi:tə]	n. 加热器，发热器
illustrate ['iləstreit]	v. 举例说明，图解
concerning [kən'sɜ:nɪŋ]	prep. 关于，就……而论
relatively ['relətɪvli]	adv. 相对地，比较地，比例地
joule [dʒu:l]	n. 焦耳
kilowatt ['kɪləwɒt]	n. 千瓦
horsepower ['hɔ:s,pauə]	n. 马力
multiply ['mʌltiplai]	v. 乘
kilowatthour ['kɪləuwɔt'auə]	n. 千瓦小时，度
extensively [ɪk'stensɪvlɪ]	adv. 广阔地
megajoule ['megdʒu:l]	n. 兆焦耳
average ['ævərɪdʒ]	adj. 平均的，普通的
summarize ['sʌməraiz]	v. 摘要，总结，概述
USCS(United States Customary System)	美国单位制
SI(System International)	国际单位制
cannot be…too strongly	无论怎样……也不算过分
in other words	换句话说
be defined as	被称为，定义为
be derived from	由……而来，从……产生
take…as	取……作为，把……看做
be divided by	被……除，以……除

⇨ Notes

[1] The fact that the watt is a unit of power or a unit of a rate of doing work cannot be emphasized too strongly.

参考译文：应该特别强调瓦特是一个功率或做功速率的单位。

由 that 引导的从句作 The fact 的同位语。

[2] Likewise, if the average power required by an electric motor for a given time is known, the energy used by the motor is the average power multiplied by the time the motor is used.

参考译文：同样，如果在给定的时间内电动机所需的平均功率是已知的，则该电动机消耗的能量就是其平均功率与电动机使用时间的乘积。

这是主从复合句。If 引导条件状语从句，其中过去分词短语 required…time 作定语，以修饰 power。

⊠ Questions

1. What is the unit of electric power?

2. What is the meaning of USCS?
3. How can we calculate the power according to formulas mentioned in this paper?

Reading Material: Electrical Instruments and Electrical Measurements

The methods of measurement of electrical quantities are so many and varied that only the more commonly used methods of measurements can be discussed here. For further details on instruments and meters, the reader is referred to the many excellent texts available on electrical measurements.

As was learned from the study of electric motors, when a current-carrying conductor is placed in a magnetic field, a force is developed on the conductor which tends to move the conductor at right angles to the field. This principle is used in current-detecting instruments. A sensitive current-detecting instrument called a galvanometer and operating on this principle is shown diagrammatically in Figure 28-1.

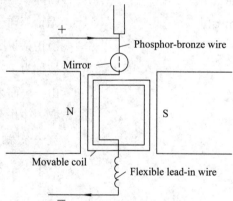

Figure 28-1 Diagram of the essentials of a galvanometer, the deflection of the movable coil is proportional to the current flowing

A coil of very fine insulated wire, usually wound on an aluminum frame or bobbin, is suspended as shown in Figure 28-1 between the poles of a permanent magnet[1]. The coil is suspended by a phosphor-bronze filament which acts as one lead-in wire for the coil. The other lead-in wire is a very flexible spiral wire at the bottom of the coil. When current flows through the coil, a deflecting force proportional to the flux density, the current, and the dimensions of the coil rotates the coil on its vertical axis. The deflecting force is opposed by the restraining force of the suspension filament so that the coil does not continue to rotate as in a motor but turns until the deflecting force is balanced by the restraining force of the suspension filament. Since the deflecting force is directly proportional to the current flowing in the coil, the amount of angular rotation may be used as an indication of the value of the current flowing in the coil.

The amount of deflection and hence the amount of current flowing through the coil may be indicated by a pointer which is attached to the moving element and moves over a calibrated scale.

On the more sensitive galvanometers a mirror is attached to the moving coil as shown in Figure 28-1. A light beam is reflected from his mirror onto a ground-glass scale. As the coil is deflected, the light beam moves over the scale.

The restoring force of the suspension filament acts to return the coil and indicating system to the normal or zero position when the current flow through the coil is interrupted.

After a reading is taken and the restoring force acts to return the coil to its normal position, the coil tends to oscillate about the normal position for some time before coming to rest. To prevent excessive oscillation of the coil, a system of damping must be employed [2]. The aluminum frame upon which the coil is wound provides the damping force in the galvanometer just described[3]. Whenever the frame moves in the magnetic field, induced current are produced which flow around the closed circuit formed by the frame. The induced currents oppose the motion that produces them.

◇ **New Words and Expressions**

varied ['veərid]	adj. 各种各样的
detail ['di:teɪl]	n. 细节，详情
angle ['æŋgl]	n. 角
principle ['prɪnsəpl]	n. 原理
detect [dɪ'tekt]	v. 探测，侦察
sensitive ['sensətɪv]	adj. 敏感的，灵敏的
wound [wu:nd]	v. 缠绕; wound (过去分词)
instrument ['ɪnstrumənt]	n. 仪器
galvanometer [ˌgælvə'nɒmɪtə]	n. 检流计
diagrammatically [ˌdaɪəgrə'mætɪkəli]	adv. 概略地，大体地
frame [freɪm]	n. 结构，框架
bobbin ['bɒbɪn]	n. 线轴，绕线筒
suspend [sə'spend]	v. 悬吊
coil [kɔɪl]	n. 线圈
filament ['fɪləmənt]	n. 细丝，游丝
flexible ['fleksəbl]	adj. 柔韧的，柔软的，能变形的
spiral ['spaɪərəl]	adj. 螺旋形的
deflect [dɪ'flekt]	v. (使) 偏斜，(使) 偏转
flux [flʌks]	n. 流动率，通量
restrain [rɪs'treɪn]	v. 抑制，约束，制止
angular ['æŋgjələ(r)]	adj. 有角的，角度测量的

pointer ['pɔɪntə(r)]	n.	指示器
attach [ə'tætʃ]	v.	附着，系上
oscillate ['ɒsɪleɪt]	v.	振荡
calibrate ['kælɪbreɪt]	v.	校准，校准……之刻度
damp [dæmp]	v.	阻尼，使衰减，抑制
standstill ['stændstɪl]	n.	停滞，停顿
tend to		倾向于，引起，有……的倾向
at right angles to...		与……成直角
phosphor bronze		磷青铜
attach to		使……连接 (安装，固定，依附)
magnetic field		磁场
act as		担任，充当，起……作用
induced current		感应电流

⇨ Notes

[1] A coil of very fine insulated wire, usually wound on an aluminum frame or bobbin, is suspended as shown in Figure 28-1 between the poples of a permanent magnet.

参考译文：一个很细的绝缘线圈通常绕在一个铝制框架或绕线筒上，如图 28-1 所示，悬吊在一个永久磁铁的两极之间。

usually wound...bobbin 过去分词短语作定语修饰 wire。

[2] To prevent excessive oscillation of the coil...

参考译文：为防止线圈过度振荡……。

不定式短语 To prevent...作主语。

[3] The aluminum frame upon which the coil is wound provides the damping force in the galvanometer just described.

参考译文：缠绕线圈的铝框就可以在刚刚讲述的检流计上提供阻尼力。

upon which...引导定语从句修饰 the aluminum frame。

⌧ Questions

1. What is the function of an electric motor?
2. How can an electric motor work?
3. Why does an electric motor need permanent magnets?

Unit 29　Numerical Control Software

Overview of Numerical Control

Today, the product design process begins with computer-generated product concepts and designs, which are subjected to detailed analysis of feasibility, manufacturability, and even disposability. Traditionally, before process planning can generate a detailed plan for manufacturing a part or assembly, design tradeoffs are made. Assemblies are broken into parts. Produced specifications are produced.

Quality criteria are determined to meet safety, environmental, and conformance with company and industry standards. Engineering drawings are produced as well as the bill of material. Final design decisions are made relating to styling, function, performance, materials, tolerances, make-versus-buy, purchased parts, supplier selection, manufacturability, quality, and reliability.

In the process of planning stage, tooling decisions are made. The sequence of production steps is planned with actions taken at each step control specified, i. e. the actions taken at each step, controls to be followed, and the state of the workpiece at each workstation. In computer-aided process planning (CAPP), an application program stores prior plans and standard sequences of manufacturing operations for families of parts coded using the group technology concept, which classifies parts based on similarity of geometric shape, manufacturing process, or some other part characteristics[1].

All tools required producing the final part or assembly are specified or designed. This includes molds, stamping and forming dies, jigs and fixtures, cut tools, and other tooling. The tool design group typically works closely with the tool room and with suppliers to produce the necessary tools in time to meet production schedules.

Numerical control provides the operational control of a machine or machines by a series of computer-coded instructions comprising numbers, letters, and other symbols, which are translated into pulses of electrical current or other output signals that activate motors or other devices to run the machine.

With NC, machines run consistently, accurately, predictably, and essentially automatically. Quality and productivity are increased, and rework and lead-times are reduced compared with manual operation APT (Automatically Programmed Tools), which was the first NC language.

ATP[2] was designed to function as an off-line, batch program using a mainframe computer system. Because of the computer resources needed and the expense, time-sharing was employed, a new simpler language to use developed. Eventually, interactive graphics-based NC programs using terminals and workstations were introduced to improve visualization and provide the opportunity for immediate feedback to the user. These developments remain the standard for NC programming today and are available on all computing platforms from PCs to mainframes.

Requirements of NC Software

NC software, accordingly, may be obtained as independent softwarepackages, as dedicated NC turnkey systems that include a printer, plotter and tape punch, or as components of CAD/CAM software packages that also provide design and drafting.

Interactive graphics-based NC part-programming technology increases product quality and simplifies the process by reducing setup time for lathes, mills, EDMs, and other machine tools. Graphics software lets users easily define part geometry, obtain immediate feedback, and visualize the results while changes are made quicky and efficiently.

NC Packages accomplish four major functions: part description, machining strategy, post-processing, and factory communications.

To describe a part, most NC Programming systems provide their own geometric modeling capability as an integral component of the system. These CAD-like frontend permits users to create parts by drawing lines, circle, arcs, and splines. All NC packages support 2D geomertry creation and many have optional 3D modules. The 3D module permits the creation of complex surfaces or direct machining from the solid model become accepted in NC.

A close relationship between the model creation and machining of that model is important. Major CAD/CAM vendors provide NC Software that is integrated with the design and drafting function for just this reason. Operating from the same or a co-existent database, the NC Software can directly access a model that has been created within the design module of the CAD/CAM system. This eliminates the requirement for translation of data from one format to another.

Other software vendors that have targeted the NC market as a specialty also include a geometric modeler as an integral component of the system. Again, a close relationship exists between a modeler and its associated NC Code generation capability. The full CAD command set can be used, there is no concern about translation of graphic data, and a single user interface is available.

Some NC Software vendors provide stand-alone systems. In this case, a file containing the definition of the model is imported from a CAD system. An IGES [3] or DXF [4] translation is usually used to convent the geometry from the originating system to the NC system. In other cases, the NC Software can directly access the CAD database and avoid translation. In dedicated N systems, complex NC calculations may be performed faster than for other systems and more NC specific utilities are often available.

Currently, NC Programming is usually done by individuals who have actual machining experience on the shop floor. This experience and the associated knowledge gained are critical in developing the machining strategy required to cut a part. It is particularly useful in handling unusual circumstances that often arise in machining complex parts.

Knowledge-based software systems can capture the NC programmer's knowledge and establish a set of rules to create a framework that can be used to lead the NC programmer through the development of the machining strategy. Default methodologies may be established based on materials, tolerances, surface finishes, machine tool availability, and part shape.

Knowledge-based systems may also be used to fully automate some NC programming tasks. This would provide for greater consistency among machining strategies and improve programming productivity.

◇ New Words and Expressions

feasibility [fiːzɪ'bɪlɪtɪ]	n. 可行性，现实性
conformance [kən'fɔːm(ə)ns]	n. 一致性，适应性
tooling ['tuːlɪŋ]	n. 工具，刀具，仪器，机床安装，工艺装备
consistency [kən'sɪst(ə)nsɪ]	n. 相容性，一致性，稳定，统一
spline [splaɪn]	n. 仿样，样条，仿样函数，样条函数
draft [drɑːft]	n. 草图，图样；v. 起草，制图
default [dɪ'fɔːlt; 'diːfɔːlt]	n. 系统设定值，隐含值，缺省值
bill of materials	材料表，材料清单
group technology	成组技术
stand-alone	独立应用
trade off	权衡，物物交换
lead time	设计(订货)至投产(交货)时间
batch program	批处理程序
dedicated turnkey system	专用的整套系统

⇨ Notes

[1] In computer-aided process...other part characteristics.
参考译文：在 CAPP 中，应用程序存储了预定的计划和加工用成组技术概念编码的零部件的标准顺序，并根据零件几何形状、加工过程以及零件的其他特征进行分类。

[2] APT：Automatically Programmed Tools. A numerical control language system used to program computers, which was first used in 1955. 用于编程的数控语言，1955 年首次应用。

[3] IGES: The Initial Graphics Exchange Specification. The standard of ANSI, which specifies the formats of geometrical information to be exchanged between different CAD systems. IGES 指的是初始图形交换规范。ANSI 标准是美国国家标准，它制定了 CAD 系统之间交换信息的格式。

[4] DXF: Drawing Interchange File. The data exchange standard of Autodesk. 图形交换

文件——Autodesk 公司的数据转换标准。

Questions

1. Why does the product design process begin with computer?
2. What is the purpose of determining quality criteria?
3. What does CAPP stand for?
4. What is the function of knowledge-based software systems?
5. What functions do NC Packages accomplish?

Reading Material: Transducers and Sensors

Analog and Digital Transducers

As mentioned previously, considerable experience has been accumulated with analog transducers[1]; signal conditioning, A/D converters[2] etc. And it is natural that the majority of current systems tend to use these techniques. However, there are a number of measuring techniques that are essentially digital in nature, and which when used as separate measuring instruments require some integral digital circuitry, such as frequency counters and timing circuits, to provide an indicator output. This type of transducer, if coupled to a computer, does not necessarily require the same amount of equipment since much of the processing done by the integral circuitry could be programmed and performed by the computer.

Collins classifies the signals handled in control and instrumentation systems as follows:

1) Analog, in which the parameter of the system to be measured although initially derived in an analog form by a sensor, is converted to an electrical analog, either by design or inherent in the methods adopted;

2) Coded-digital, in which a parallel digital signal is generated, each bit radix—weighted according to some predetermined code. These are referred to in this book as direct digital transducers;

3) Digital, in which a function, such as mean rate of a repetitive signal, is a measure of the parameter being measured. These are subsequently referred to as frequency-domain transducers.

Some analog transducers are particularly suited to conversion to digital outputs using special techniques. The most popular of these are synchros, and similar devices which produce a modulated output of a carrier frequency. For ordinary analog use, this output has to be demodulated to provide a dc signal whose magnitude and sign represents any displacement of the transducer's moving element. Although it is possible to use a conventional A/D technique to produce a digital output, there are techniques by which the synchro output can be converted directly to a digital output while providing a high accuracy and resolution, and at a faster rate than is possible in the A/D converter method.

Direct digital transducers are, in fact, few and far between, since there do not seem to be any natural phenomena in which some detectable characteristic changes in discrete intervals as a result of a change of pressure, or change of temperature etc[3]. There are many advantages in using direct digital transducers in ordinary instrumentation systems, even if computers are not used in the complete installation. These advantages are:

(a) The ease of generating, manipulating and storing digital signals, as punched tape, magnetic tape etc;

(b) The need for high measurement accuracy and discrimination;

(c) The relative immunity of a high-level digital signal to external disturbances (noise);

(d) Ergonomic advantages in simplified data presentation (e.g. digital readout avoids interpretation errors in reading scales or graphs);

(e) Logistic advantage concerning maintenance and spares compared with analog or hybrid systems.

The most active development in direct digital transducers has been in shaft encoders, which are used extensively in machine tools and in aircraft systems. High resolution and accuracies can be obtained, and these devices may be mechanically coupled to provide a direct digital output of any parameter which gives rise to a measurable physical displacement. For example, a shaft encode attached to the output shaft of a Bourdon tube gauge can be used for direct pressure measurement of temperature measurement using vapour pressure thermometers. The usual disadvantage of these systems is that the inertia of the instrument and encoder often limits the speed of response and therefore the operating frequencies.

Frequency domain transducers have a special part to play in on-line systems with only few variables to be measured, since the computer can act as part of an A/D conversion system and use its own registers and clock for counting pulses or measuring pulse width[4]. In designing such systems, consideration must be given to the computer time required to access and process the transducer output.

Use of Sensors in Programmable Automation

In this paper we are concerned with the application of sensor-mediated programmable automation to material-handling, inspection, and assembly operation in batch-produced, discrete-part manufacturing.

Programmable automation consists of a system of multidegree-of-freedom manipulators[5] (commonly known as industrial robots[6]) and sensors under computer control, which can be programmed to perform specified jobs in the manufacturing process and can be applied to new (but similar) jobs by reprogramming[7]. This is particularly important where production runs are small and where different models may have to be produced frequently. Today industrial robots have neither contact sensors as aids to manipulation, nor noncontact sensors as aids to recognition, inspection, or manipulation of workpiece.

Extending the present capabilities of industrial robots will require a considerable improvement in their capacity to perceive and interact with the surrounding environment. In particular, it is desirable to develop sensor-mediated, computer-controlled interpretive systems that can emulate human capabilities. To be acceptable by industry, these hardware/software systems must perform as well or better than human workers. Specifically, they must be inexpensive, fast, reliable, and suitable for the factory environment.

Sensors can be broadly divided into three areas of application: visual inspection, finding parts, and controlling manipulation.

Visual inspection. Here we concerned only with an important aspect of visual inspection: the qualitative and semiquantitative type of inspection performed by human vision rather than by measuring instruments. Such inspection of parts or assemblies includes identifying parts; detecting of burrs, cracks, and voids; examing cosmetic qualities and surface finish; counting the number of holes and determing their locations and sizes; assessing completeness of assembly and so on. It is evident that a large library of computer programs will have to be developed to cope with the numerous classes of inspection.

Finding parts. For material-handling and assembly operations in the unstructured environment of great majority of factories, it will probably be necessary to "find" work pieces—that is, to determine their positions and orientations and sometimes also to identify them. Thus it is necessary to augment existing robots with visual sensors to be able to determine the identity, position, and orientation of parts and to perform visual inspection.

Controlling manipulation. It appears useful to consider the use of both contact and noncontact sensors in manipulator control and to try assessing where each sensor is most appropriate. One approach is to divide the sensory domain into coarse and fine sensing, using noncontact sensors for coarse resolution and contact sensors for fine resolution. For example, in acquiring a work piece that may be randomly positioned and oriented, a visual sensor may be used to determine the relative position and orientation of the workpiece rather coarsely, say, to one tenth of an inch. From this information the manipulator can be positioned automatically. The somewhat compliant fingers of the manipulator hand, bracketing the workpiece, will now be close enough to effect closure, relying on touch sensors to stop the motion of each finger when a specified contact pressure is detected. After contacting the workpiece without moving it, the compliant fingers have flexed no more than a few thousandths of an inch before a stopping. The touch sensors have thus performed fine resolution sensing and have compensated for the lack of precision of both the visual sensor and the manipulator. It is a quite common task to illustrate the relative merits of each sensory modality and the advantage of using both.

Other common applications for contact sensors, which entail fine resolution or precision sensing, include:

Collision avoidance, using force sensors on the links and hand of a manipulator. Motion is quickly stopped when any one of preset force thresholds is exceeded.

Packing operation, in which parts are packed in orderly fashion in tote boxes. Force sensors can be used to stop the manipulator when its compliantly mounted hand touches the bottom of the box, its sides, or neighboring parts. This mode of force feedback compensates for the variability of the positions of the box and the parts and for the small but important variability of the manipulator positioning.

Insertions of pegs, shafts, screws and bolts into holes. Force and torque sensors can provide feedback information to correct the error of a computer-controlled manipulator.

◆ *New Words and Expressions*

analog ['ænəlɒg]	n. 模拟(量，装置，设备，系统，计算机)
transducer [trænz'dusɚ]	n. 变(转)换器，传感器
converter [kən'vɜːtə]	n. 转换器
counter ['kaʊntə]	n. 计数器
couple ['kʌp(ə)l]	v. 耦[联，结，组，配]合
instrumentation [,ɪnstrʊmen'teɪʃ(ə)n]	n. (检测)仪(器，表)，测试设备
initially [ɪ'nɪʃ(ə)lɪ]	adv. 最[起]初，开头，一开始
derive [dɪ'raɪv]	v. (从……)得到(出)，取(获)得，导(引，伸)出
sensor ['sensə]	n. 传感器，传感[敏感]元件，探测器
code [kəʊd]	v. 编码，代码
function ['fʌŋ(k)ʃ(ə)n]	n. 函数
synchro ['sɪŋkrəʊ]	n. (自动)同步机(器)
modulate ['mɒdjʊleɪt]	v. 调(制，整，节，幅，谐)
displacement [dɪs'pleɪsm(ə)nt]	n. 位移，平移
resolution [rezə'luːʃ(ə)n]	n. 分辨率
phenomena [fə'nɒmɪnə]	n. 现象，征兆
discrete [dɪ'skriːt]	adj. 不连续的，离散的，分立的
interval ['ɪntəv(ə)l]	n. 间隔(距，隙)，时间间隔
manipulate [mə'nɪpjʊleɪt]	v. 操作(纵)，控制，管理，处理
discrimination [dɪ,skrɪmɪ'neɪʃ(ə)n]	n. 辨(鉴，区，判，识)别，区分，辨别力
immunity [ɪ'mjuːnɪtɪ]	n. 抗扰性，免疫(性，力)
ergonomic [,ɜːgəʊ'nɒmɪk]	adj. 人机工程学的
encoder [en'kəʊdə]	n. 编[译]码器，编码装置，编码员
access ['ækses]	v. 存取，取款，(数据，信息)选取，访问
batch [bætʃ]	n. 一批
mediate ['miːdɪeɪt]	v. 处于中间，介乎其间，作为引起……的媒介，传递
manipulator [mə'nɪpjʊleɪtə(r)]	n. 机械手
robot ['rəʊbɒt]	n. 机器人
recognition [rekəg'nɪʃ(ə)n]	n. 识别，认出，辨别

perceive [pə'siːv]	v. 察觉，发觉，看见，看出
interpretive [ɪn'tɜːprɪtɪv]	adj. 解释 (翻译，说明) 的
emulate ['emjʊleɪt]	v. 模仿 [拟]，仿真
reliable [rɪ'laɪəb(ə)l]	adj. 可靠的
visual ['vɪʒʊəl]	adj. 视觉的，可见的
burr [bɜː]	n. 毛口 (刺，边)
void [vɒɪd]	n. 空洞 (穴)
orientation [ɔrɪɛn'teʃən]	n. 定 (取) 向，定 (方位)，方位
augment [ɔːg'ment]	v. 增大 (加，长)，添增
randomly ['rændəmli]	adv. 随机地，无规律地
compliant [kəm'plaɪənt]	adj. 依从的，顺从的，屈从的
flex [fleks]	v. (使) 弯曲，挠 (屈) 曲
compensate ['kɒmpenseɪt]	v. 补偿 (助，充，整)
modality [mə(ʊ)'dælɪtɪ]	n. 模态，形态
entail [ɪn'tel]	v. 需要，要求，使……发生
threshold ['θrɛʃhold]	n. 门槛，阈值，门限值，限度
peg [peg]	n. 柱 (钉)，销
analog and digital converter	模/数 (A/D) 转换器
signal conditioning	信号修整
radix-weighted	基数加权
integral-circuit	积分数字电路
timing-circuit	计时电路
frequency-domain	频域转换器
mean-rate-of-repetitive	重复信号的平均率
carrier-frequency	载 (波) 频 (率)
few-and-far-between	极少，稀少
burdon-tube-gauge	布尔登式压力计
multidegree-of-freedom	多自由度
tote-box	运输斗

⇨ Notes

[1] transducer: Any device or element which converts an input signal into an output signal of a different form. An example is the microphone, which converts vibration caused by an impinging sound wave into an electrical signal. 转换器：能将输入信号转换成不同形式的输出信号的设备或元件，典型的转换器是麦克风，它能将传来的声波引起的振动变成电信号。

[2] A/D converter (Analog-to-Digital Converter): A device for converting the information contained in the value or magnitude of some characteristic of an input signal, compared to a

standard or reference, to information in the form of discrete states of a signal, usually with numerical values assigned to the various combinations of discrete states of the signal. A/D 转换器 (模/数转换器)：将依据某标准或参考系得到的输入信号中的某些特性值所含的信息转变成信号的离散状态形式的信息的设备，通常是对信号的不同的离散状态赋以的数值。

[3] Direct digital transducers are… or change of temperature etc.
参考译文：直接数字转换器事实上是极少的，因此似乎还没有任何自然现象，它的可检测到的特性变化，如压力变化、温度变化等是离散变化的。

[4] Frequency domain transducers… measuring pulse width.
参考译文：频域转换器在只有几个待测变量的在线系统中有着重要的作用，因为计算机可以起到一部分 A/D 转换系统的作用，使用它自己的寄存器及时钟来对脉冲计数或测定脉冲宽带。

[5] manipulator：A mechanical device for handling objects as desired without touching them with the hands. 机械手：按人类的意愿处理物体的机械设备，不必用手接触物件。

[6] industrial robot: A programmable mechanism designed to move and do work with a certain volume of space. The robot differs from a parts-transfer mechanism in that the action patterns can easily be changed by software changes in a controlling computer or sometimes by adjusting the mechanism. 工业机器人：可编程机构，设计用来在一定空间内移动或处理工件。机器人与零件传输机构的区别在于机器人的动作方式很容易通过调整机构来实现。

[7] Programmable automation consists…jobs by reprogramming.
参考译文：可编程自动化是几个多自由度机械手(通常称作工业机器人)和计算机控制下的传感器组成的一个系统，可对它进行编程以进行加工中的专门工作，也可通过再编程使之执行新的 (但是类似的) 工作。

⊠ Questions

1. How does Collins classify the signals handled in control and instrumentation systems?
2. What are advantages of using direct digital transducers in ordinary instrumentation systems?
3. What does programmable automation consist of?
4. What should be paid attention to when using sensors in programmable automation?
5. When can force sensors be used to stop the manipulator?

Unit 30 Automatic Control System

Introduction

In recent years, automatic control systems have assumed an increasingly important role in the development and advancement of modern civilization and technology. Domestically automatic controls in heating and air conditioning systems regulate the temperature and the humidity of modern homes for comfortable living. Industrially, automatic control systems are found in numerous applications, such as quality control of manufactured products, automation, machine tool control, modern space technology and robotics[1]. Even such problem as inventory control, social and economic systems control may be approached from the theory of automatic control.

The simple block diagram shown in Figure 30-1 may describe the basic control system concept. The objective of the system is to control the variable in a prescribed manner by the actuating signal[2] through the elements of the control system.

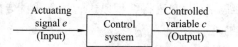

Figure 30-1 Basic control system

In more common terms, the controlled variable is the output of the system, and the actuating signal is the input. As a simple example, in the steering control of an automobile, the direction of the two front wheels may be regarded as the controlled variable, the output, the position of the steering wheel is the input, the steering mechanisms, including the dynamics of the entire automobile.

However, if the objective is to control the speed of the automobile, then the amount of pressure exerted on the accelerator is the actuating signal, with the speed regarded as the controlled variable.

There are many situations where several variables are to be controlled simultaneously by a number of inputs; such systems are referred to as multivariable systems.

Open-loop Control System

The word automatic implies that there is a certain amount of sophistication in the control system. By automatic, it generally means that the system is usually capable of adapting to a variety of operating conditions and is able to respond to a class of inputs satisfactorily[3]. However, not any type of control system has the automatic feature. Usually, the automatic feature is achieved by feeding the output variable back and comparing it with the command signal. When a system does not have the feedback[4] structure, it is called an open-loop system, which is the simplest and most economical type of control system. Unfortunately, open-loop control systems[5] lack accuracy and versatility and can be used in none but the simplest types of applications.

Consider, for example, control of the furnace for home heating. Let us assume that the furnace is equipped with a timing device, which controls the on and off periods of the furnace. To regulate the temperature to the proper level, the human operator must estimate the amount of time required for the furnace to stay on and then set the timer accordingly. When the present time is up the furnace is turned off. However, it is quite likely that the house temperature is either above or below the desired value, owing to inaccuracy in the estimate. Without further deliberation, it is quite apparent that this type of control is inaccurate and unreliable. One reason for the inaccuracy lies in the fat that one may not know the exact characteristics of the furnace. The other factor is that one has no control over the outdoor temperature, which has a definite bearing on the indoor temperature. This also points to an important disadvantage of the performance of an open-loop control system, in that the system is not capable of adapting to variations in environmental conditions or to external disturbances. In the case of the furnace control, perhaps an experienced person can provide control for a certain desired temperature in house; but if the doors or windows are opened or closed intermittently during the operating period, the final temperature inside the house will not be accurately regulated by the open-loop control.

An electric washing machine is another typical example of an open-loop system, because the amount of wash time is entirely determined by the judgment and estimation of the human operator. A true automatic electric washing machine should have the means of checking the cleanliness of the clothes continuously and turn itself off when the desired degree of cleanness is reached.

Although open-loop control systems are of limited use, they form the basic elements of the closed-loop control systems[6]. In general, the elements of an open-loop control system are represented by the block diagram of Figure 30-2. An input signal or command is applied to the controller, whose output acts as the actuating signal; the actuating signal then actuates the controlled process and hopefully will drive the controlled variable to the desired value.

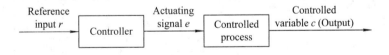

Figure 30-2 Block diagram of an open-loop control system

Closed-loop Control System

What is missing in the open-loop control system for more accurate and more adaptable control is a link or feedback from the output to input of the system. In order to obtain more accurate control, the controlled signal must be fed back and compared with the reference input, and an actuating signal proportional to the difference of the output and the input must be sent through the system to correct the error[7]. A system with one of more feedback paths like that just described is called a closed-loop system. Human beings are probably the most complex and sophisticated feedback control system in existence. A human being may be considered to be a control system with many inputs and outputs, capable of carrying out highly complex operations.

To illustrate the human being as a feedback control system, let us consider that the objective is to reach for an object on a desk. As one is reaching for the object, the brain sends out a signal to the arm to perform the task. The eyes serve as a sensing device, which feeds back continuously the position of the hand. The distance between the hand and the object is the error, which is eventually brought to zero as the hand reached the object. This is a typical example of closed-loop control. However, if one is told to reach for the object and then is blindfolded, one only reach toward the object by estimating its exact position. It is quite possible that the object may be missed by a wide margin. With the eyes blindfolded, the feedback path is broken, and the human is operating as an open-loop system.

As another illustrative example of a closed-loop control system. Figure 30-3 shows the block diagram of the rudder control system of a ship. In this case the objective of control is the position of the rudder, and the reference input is applied through the steering wheel. The error between the steering wheel and the rudder is the signal, which actuates the controller and the motor. When the rudder is finally aligned with the desired reference direction, the output of the error sensor[8] is zero. Let us assume that the steering wheel position is given a sudden rotation of R units, as shown by the time signal in Figure 30-4(a). The position of the rudder as a function of time, depending upon the characteristics of the system, may typically be one of the responses[9] shown in Figure 30-4(b). Because all physical systems have electric and mechanical inertia, the position of the rudder cannot respond instantaneously to step input, but will, rather move gradually toward the final desired position. Often, the response will oscillate about the final position before settling. It is apparent that for the rudder control it is desirable to have a nonoscillatory response.

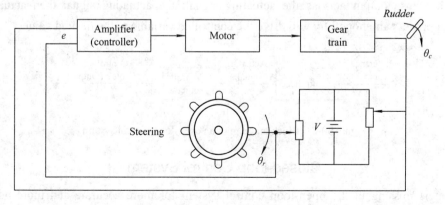

Figure 30-3   Rudder control system

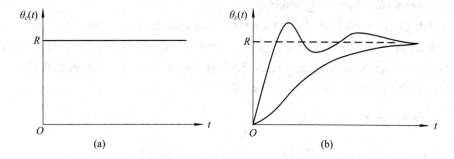

Figure 30-4 (a) Step displacement input of rudder control system; (b) Typical output responses

The basic elements and the block diagram of a closed-loop control system are shown in Figure 30-5. In general, the configuration of a feedback control system may not be constrained to that of Figure 30-5. In complex systems there may be a multitude of feedback loops and element blocks.

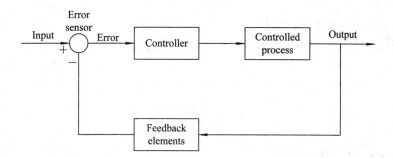

Figure 30-5 Basic elements of feedback control system

Figure 30-4(a) illustrates the elements of a tension control system of a windup process. The unwind reel may contain a roll of material such as paper or cable which is to be sent into a processing unit such as a cutter or a printer, and then collects it by winding it onto another roll. The control or a printer, and then collects it by winding it onto another roll. The control system in case is to maintain the tension of the material or web[10] at a certain prescribed tension to avoid such problems as tearing, stretching, or creasing.

To regulate the tension, the web is formed into a half-loop by passing it down and around a weighted roller. The roller is attached to a pivot arm, which allows free up-and-down motion of the roller. The combination of the roller and the pivot arm is called the dancer[11].

When the system is in operation, the web normally travels at a constant speed. The ideal position of the dancer is horizontal, producing a web tension equal to one-half of the total weight of the dancer roll. The electric brake on the unwind reel is to generate a restraining torque to keep the dancer in the horizontal position at all times.

During actual operation, because of external disturbances, uncertainties and irregularities of the web material, and the decrease of the effective diameter of the unwind reel, the dancer arm will not remain horizontal unless some scheme is employed to properly sense the dancer-arm position and control the restraining braking torque.

To obtain the correction of the dancing-arm-position error, an angular sensor is used to measure the angular deviation, and a signal in proportion to the error is used to control the braking torque through a controller. Figure 30-4(b) shows a block diagram that illustrates the interconnections between the elements of the system.

◇ *New Words and Expressions*

hydrological [ˌhaɪdrə'lɑdʒɪkəl]	adj. 水文 (学) 的
inventory ['ɪnv(ə)nt(ə)rɪ]	n. 库存量，存货 (清单)
approach [ə'prəʊtʃ]	n. 研究，接近，处理
variable ['veərɪəb(ə)l]	n. (可) 变量，(可) 变 (因) 素
steer [stɪə]	v. 驾驶，操作(向)，掌舵
simultaneously [ˌsɪml'teɪnɪəslɪ]	adv. 同时地，同步地，一齐地
versatility [ˌvɜːsə'tɪlətɪ]	n. 通用性，多功能性，各方面的适用性
disturbance [dɪ'stɜːb(ə)ns]	v. 扰动 (乱)，干扰
intermittently [ɪntə'mɪtəntlɪ]	adv. 间歇 (断) 地，周期性地
regulate ['regjʊleɪt]	v. 调整 (节，准)，校 (对) 准，控制
signal ['sɪgn(ə)l]	n. 信号
feedback ['fiːdbæk]	n. 反馈
error ['erə]	n. 误差
blindfold ['blaɪn(d)fəʊld]	v. 蒙住……的眼睛，遮住……的视线
align [ə'laɪn]	v. 直线对 [照] 准，定位，对中
rudder ['rʌdə]	v. (方向) 舵，舵手
inertia [ɪ'nɜːʃə]	n. 惯性 (物)，惯量，惰性
oscillate ['ɒsɪleɪt]	v. 振荡，摇摆
response [rɪ'spɒns]	n. 应答，回答，响应
settle ['set(ə)l]	v. 稳 (安) 定，决 [确] 定
configuration [kənˌfɪgjə'reʃən]	n. 结构，组合 (态)，布置，(设备) 配置
tension ['tenʃ(ə)n]	n. 张 (拉，牵，弹) 力，拉伸
reel [riːl]	n. 卷轴，线轴，卷筒，绕线筒 (架，管)
web [web]	n. (印报纸的) 卷筒纸
tear [teə]	v. 撕开 (裂)，扯开
stretch [stretʃ]	v. 伸 (展，长)，拉 (直，长，紧)
crease [kriːs]	v. (使) 起折痕，变皱
dancer ['dɑːnsə]	v. (松紧) 调节器，浮动滚筒
brake [breɪk]	n. 制动器，闸，刹车
torque [tɔːk]	n. 转矩，扭矩
deviation [diːvɪ'eɪʃ(ə)n]	n. 偏差

actuating signal	动作 (起动，伺服) 信号
open-loop control system	开环控制系统
closed-loop control system	闭环控制系统
step input	步进输入
a multitude of	许多的，众多的
block diagram	框图

⇨ Notes

[1] robotics: a field of interest concerned with the construction, maintenance, and behavior of robots. 机器人学：有关机器人的生产、维修及其运转状态的研究领域。

[2] signal: a detectable physical quantity or impulse (as a voltage, current, or magnetic field strength) by which messages or information can be transmitted. 信号：可检测的物理量或脉冲，如电压、电流、磁场强度，通过该量可传送信息。

[3] The word automatic implies… to a class of inputs satisfactorily.
参考译文：自动化的意思是在控制系统中一定程度地采用了先进技术。所谓自动化，一般是指系统通常能适应各种操作条件，能令人满意地对一类输入做出响应。

[4] feedback: the return to the input of a part of the output of a machine, system, or process (as for producing changes in an electronic circuit that improve performance or in an automatic control device that provide self-corrective action). 反馈：将机器、系统或过程中的输出的一部分返回到输入中去，如：可在电子电路中产生一个改变以改善性能或在自动控制设备中提供自动修正作用。

[5] open-loop control system: a control system for an operation or process in which there is no self-corrective action as there is in a closed loop. 开环控制系统：用于操作或过程的一种控制系统，该系统同闭环系统不同，没有自我修正作用。

[6] closed-loop control system: an automatic control system for an operation or process in which feedback in a closed path or group of paths acts to maintain output at a desired level. 闭环控制系统：一种用于操作或过程的自动控制系统，在该系统中，一条或几条闭合回路中的反馈起作用，以便输出保持在一定的水平上。

[7] In order to obtain… to correct the error.
参考译文：为了获得更精确的控制，必须要将控制信号反馈回来，与标准输入信号进行比较，然后向系统发出一个与输出、输入信号之差成比例的伺服信号以修正误差。

[8] sensor：A device that responds to a physical stimulus (as heat, light, sound, pressure, magnetism, or a particular motion), and transmits a resulting impulse (as for measurement or operating a control). 传感器：能对物理激励源，如热、光、声、压力、磁或某一特殊运动产生响应，并能传送一个由此产生的脉冲以作测量或控制用的一种设备。

[9] response: The output of a transducer or detecting device resulting form a given input. 响应：传感器或检测设备对某一输入产生的输出。

[10] web: a roll of paper for use in a rotary printing press. 卷筒纸：在滚筒式印刷机上的

一卷纸。

[11] To regulate the tension, …is called the dancer.

参考译文：为了调节张力，卷筒纸向下绕过一个挂着重物的滚筒，滚筒与一转轴相连，从而可使滚筒上下自由运动，滚筒与转轴就组成松紧调节辊。

Questions

1. In recent years, why have automatic control systems became an increasingly important role in modern civilization and technology?
2. What is the difference between open-loop control system and closed-loop control system?
3. What is the function of open-loop control system?
4. What are advantages of closed-loop control system?
5. What are defects of open-loop control system?

Reading Material: Artificial Intelligence for Automotive Manufacturing

Applications

In an effort to continue to meet global competition, the automotive industry is incorporating artificial intelligence and expert systems into their manufacturing plants. For instance, GM is taking a heuristic-based approach to develop an expert system for production scheduling a wide range of parts, from axles to ignition systems that can vary in quantity from 6 to 500 parts/tray[1]. The product objectives include the development of a generic scheduler that can be used at several sites for production involving days or weeks. In other words, the system is designed for a flexible factory environment that involves repetitive discrete manufacturing and/or batch production.

The scheduler itself must be able to meet customer orders in a cost effective way and be able to observe operating constrains. It must also be able to choose the best resource for each event, depending on current conditions and provide a unique scheduler of resource utilization as well. Inputs to the scheduler include factory conditions, expected receipts, and customer part orders, service orders, preventive maintenance requirements, inventory, labor data, tooling information, and purchased parts, among others. The output generated by the scheduler includes parts flow schedules, preventive maintenance schedules, labor/tooling requirements, shipping schedules, and short-term resource requirements. These outputs are sent to the factory control system, which has multi-task capabilities.

Though GM's system still has problems with data consistency and merging of new/existing schedules, it is being used for schedules of parts flows at GM's Saginaw as of September 1988 and as a planning tool at GM's Allison transmission division. The project plans to enhance the heuristics approach, by using simulation to compare results and will eventually use the scheduling system for planning labor and tooling requirements.

Ford Motor Company is also using AI and expert system for manufacturing and engineering as well. Typical applications include design of fuel devices, fastener selection, and electronic component selection. In the area of manufacturing, applications include process planning, generation of work standards and time, and production scheduling. Ford is also looking at expert systems for machine diagnostics to assist in repair and maintenance, as well as vehicle diagnostics for sales and service departments and as training aids. Ford's ultimate goal is to solve real problems in real time, no matter how small, whether it is in the engineering design or manufacturing environment[2]. This will require a user-friendly system, general purpose hardware, such as personal computers, database interfaces, a distributed database, and parallel processing.

Research

Purdue University's center for intelligent systems for manufacturing is helping companies like GM and Ford achieve their goals. The center is looking at integration of artificial intelligence with manufacturing systems that are fully integrated, flexible, adaptive, and compute controlled. An interdisciplinary approach is being used that involves artificial intelligence, automatic control robotics, dexterous manipulation, sensing, networking, modeling, and simulation. Application being investigated includes customer interfacing, design, production planning, and material handling. The center is also developing a robotic scanner incorporation fiber optics that can learn the shape of a part by repetitive scanning.

On the other hand, researchers at Rensselaer Polytechnic University are taking a different approach to artificial intelligence. Their methodology relies on symbolic engineering models instead of expert systems. This model approach must address five issues concurrently for optimum knowledge system development: purpose, representation, reasoning, interfacing user with the system, and testing. The objective is to develop a system with reusable application modules that can incorporate useful new technology.

For successful implementation, the following is required: realistic expectations, appropriate tools and services, management backing, and well-chosen problems. Rensselaer Polytechnic University has applied their modeling approach to the production scheduling of semiconductor wafers at a plant that fabricates 10 different parts using 250 machines. The entire factory must be modeled as well as the parts and process plans. The system is intended for such user as planners and shift supervisors. Performance is reported as actual delivery of parts against required schedule.

◇ New Words and Expressions

heuristic [ˌhjʊ(ə)'rɪstɪk]   adj. 启发式的，探索的，渐进的
ignition [ɪg'nɪʃ(ə)n]   n. 点火，引燃

tray [treɪ]	n. 托盘，支架，垫，座
generic [dʒɪ'nerɪk]	adj. 一般的，通用的，非特有的
scheduler ['ʃedjuːələ]	n. 程序机，计划安排
dexterous ['dekst(ə)rəs]	adj. 灵巧的，惯用右手的
wafer ['weɪfə]	n. 片，晶片，垫片
shipping ['ʃɪpɪŋ]	n. 发货，海运
transmission [trænsˈmɪʃən]	n. 传递，传动装置，变速器
artificial intelligence	人工智能
preventive maintenance	预防性维修
robotic scanner	自动扫描仪
look at	考虑

⇨ **Notes**

[1] tray: The work to be done by one working group in the same time with only one drawing, the same equipment and tools. 托盘：由一个工作组在同一时间、同一图纸，使用同一设备和工具所完成的工作。

[2] Ford's ultimate goat is to solve real problems in real time, no matter how small, whether it is in the engineering design or manufacturing environment.

参考译文：福特的最终目的是：无论实际问题有多小，也不管它是出现于手工程序设计中或加工过程中，都要在第一时间得到解决。

⊠ **Questions**

1. Why is the automotive industry incorporating artificial intelligence into manufacturing plants?
2. What is GM doing in the field of artificial intelligence?
3. What objectives do the GM products have?
4. What problems does GM's system still have?
5. What is Purdue University's center for intelligent systems for manufacturing responsible for?

参 考 文 献

[1] 美国麦克米伦《职业英语》编委会. 土木和机械工程英语[M]. 北京：世界图书出版公司，1991.
[2] 贺自强，等. 机械工程专业英语[M]. 北京：北京理工大学出版社，1989.
[3] 戴浩中. 机械英语自学读本[M]. 上海：上海科学技术出版社，1981.
[4] 周菊琪，徐旭东. 机械制造工程专业英语[M]. 北京：学苑出版社，1993.
[5] 伍忠杰. 机械专业英语[M]. 北京：北京理工大学出版社，1996.
[6] 李万莉等. 工程机械专业英语[M]. 北京：人民交通出版社，1998.
[7] 施平. 机械工程专业英语[M]. 哈尔滨：哈尔滨工业大学出版社，1999.
[8] (英)格兰迪. 牛津电气与机械工程学[M]. 北京：北京大学出版社，1998.
[9] 司徒忠，李璨. 机械工程专业英语[M]. 武汉：武汉理工大学出版社，2001.
[10] 陈统坚. 机械工程英语[M]. 北京：机械工业出版社，1996.
[11] 黄运尧，司徒忠. 机械类专业英语阅读教程[M]. 北京：机械工业出版社，1997.
[12] 章跃. 机械制造工程专业英语[M]. 北京：机械工业出版社，2004.
[13] 章跃. 机械制造专业英语[M]. 北京：机械工业出版社，2003.
[14] 董建国. 机械专业英语[M]. 西安：西安电子科技大学出版社，2004.
[15] 杨正. 机械工程专业英语[M]. 北京：中国电力出版社，2004.
[16] 施平主. 机械工程专业英语教程[M]. 北京：电工电子出版社，2003.
[17] 赵红霞，王淑. 机械工程专业英语[M]. 武汉：武汉理工大学出版社，2005.
[18] 施平主. 机械工程专业英语教程[M]. 北京：电子工业出版社，2003.
[19] 赵运才，何法江. 机电工程专业英语[M]. 北京：北京大学出版社，2006.
[20] 程安宁，周新建. 机械工程科技英语[M]. 西安：西安电子科技大学出版社，2007.
[21] 宋红英. 汽车专业英语[M]. 北京：机械工业出版社，2003.
[22] 黄汽驰. 汽车专业英语[M]. 北京：机械工业出版社，2005.
[23] 郑殿旺，陈庆新. 汽车英语阅读[M]. 哈尔滨：哈尔滨工业大学出版社，1998.
[24] 宋进桂. 汽车专业英语读译教程[M]. 北京：机械工业出版社，2007.
[25] 冯春燕. 通信与计算机专业英语[M]. 北京：电子工业出版社，1998.
[26] 周世麟，董恒. 铸造专业英语文选[M]. 北京：机械工业出版社，1989.
[27] 许庆衍. 化学化工专业英语[M]. 北京：中国物资出版社，1994.
[28] 李万莉. 工程机械专业英语[M]. 北京：人民交通出版社，1998.
[29] 谈振藩. 自动控制专业英语[M]. 哈尔滨：哈尔滨工业大学出版社，1999.
[30] 王开铸. 计算机专业英语阅读[M]. 哈尔滨：哈尔滨工业大学出版社，1997.
[31] 何人可. 工业设计专业英语[M]. 北京：北京理工大学出版社，1999.
[32] 李洪涛，费维栋. 材料科学英语阅读[M]. 哈尔滨：哈尔滨工业大学出版社，1999.

[33] 周光垌. 力学与工程科学专业英语 [M]. 北京：机械工业出版社，1997.

[34] 赵淑清. 电子信息与通信专业英语 [M]. 哈尔滨：哈尔滨工业大学出版社，2000.

[35] 李久胜. 电气自动化专业英语 [M]. 哈尔滨：哈尔滨工业大学出版社，1999.

[36] 刘然，包兰宇，景志华. 电力专业英语 [M]. 北京：中国电力出版社，2004.

[37] 张军. 材料专业英语：译写教程 [M]. 北京：机械工业出版社，2001.

[38] 陈焕江，徐双应. 交通运输专业英语 [M]. 北京：机械工业出版社，2002.

[39] 庄佩君. 物流专业英语 [M]. 北京：电工工业出版社，2004.

[40] 张宝华. 造船专业英语 [M]. 哈尔滨：哈尔滨工业大学出版社，2003.